Alcohol
Interactions
with
Drugs
and
Chemicals

Edward J. Calabrese

CRC Press
Taylor & Francis Group
Boca Raton London New York

CRC Press is an imprint of the
Taylor & Francis Group, an **informa** business

Edward J. Calabrese is a board certified toxicologist who is professor of toxicology at the University of Massachusetts School of Public Health, Amherst. Dr. Calabrese has researched extensively in the area of host factors affecting susceptibility to pollutants, and is the author of more than 270 papers in scholarly journals, as well as 13 books, including *Principles of Animal Extrapolation, Nutrition and Environmental Health,* Vols. I and II, *Ecogenetics, Safe Drinking Water Act: Amendments, Regulations and Standards, Petroleum Contaminated Soils,* Vols. 1, 2, and 3, *Ozone Risk Communication and Management, Hydrocarbon Contaminated Soils, Volume 1,* and *Hydrocarbon Contaminated Soils and Groundwater, Volume 1.* His most recent books include *Multiple Chemical Interactions* and *Air Toxics and Risk Assessment.* He has been a member of the U.S. National Academy of Sciences and NATO Countries Safe Drinking Water committees, and most recently has been appointed to the Board of Scientific Counselors for the Agency for Toxic Substances and Disease Registry (ATSDR). Dr. Calabrese also serves as Chairman of the International Society of Regulatory Toxicology and Pharmacology's Council for Health and Environmental Safety of Soils (CHESS) and Director of the Northeast Regional Environmental Public Health Center at the University of Massachusetts.

PREFACE

The most significant drug of abuse in the United States is alcohol. It is responsible for thousands of deaths in automobile accidents, numerous social problems, and countless stories of personal tragedy. Despite the high visibility that alcohol abuse has within society, it is generally not recognized that alcohol consumption may play a critical role in affecting susceptibility to a wide range of environmental and industrial contaminants. This book was designed to identify, document, and assess the capacity of alcohol to alter the toxicity of pollutants and drugs in animal models and humans.

The results of this study have yielded surprising insights on the significance of alcohol consumption as a major factor affecting the impact of environmental pollutants on human health. While all interactions between alcohol and pollutants and drugs will certainly not be harmful, many are. This fact is not only in need of recognition but emphasis as well.

The general myth that this book dissolves is that only heavy drinkers or alcoholics need be concerned with the interaction of ethanol and pollutants and drugs. For example, as little as a single beer can alter the metabolism of highly carcinogenic nitrosamines and enhance cancer risk.

This book is organized to systematically assess interactions according to general chemical classes, with inorganic agents being addressed in Section II while a wide range of organic agents including numerous drugs and pollutants are presented in Section III. The book concludes with an integrative discussion of the public health significance of this area (Section IV).

This book will be of interest to a wide variety of scientific disciplines including environmental scientists, toxicologists, epidemiologists, and alcohol researchers. This information will be helpful in establishing productive interdisciplinary relationships of professionals in these various areas to more effectively establish ways in which to move this exciting and critical field toward better insights as to the role of alcohol in affecting drug and pollutant interactions.

TABLE OF CONTENTS

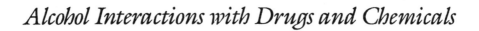

Alcohol Interactions with Drugs and Chemicals

ALCOHOL INTERACTIONS WITH DRUGS AND CHEMICALS

I. INTRODUCTION

Perhaps the most pervasive agent consumed in substantial quantities that affects that toxicity of dozens of toxic substances is ethanol. Consumption of ethanol has been shown to affect the pharmacokinetics of numerous agents via multiple mechanisms. This chapter identifies nearly 40 agents whose toxicity is modified by the presence of ethanol. It is likely that this rather large number of agents is merely the tip of a very large iceberg. This chapter will identify and critique the literature on ethanol and xenobiotic interaction for both inorganic and organic agents, with particular emphasis on the elucidation of biochemical mechanisms and predictive model development as well as the public health implications of the information.

II. INORGANICS

A. Cadmium

Since the administration of ethanol and cadmium cause hepatotoxic effects, Kershaw et al. (1988) assessed the interaction of these two agents. More specifically, rats (strain and sex not given) were orally administered 7 g of ethanol in 10 mL H_2O/kg 24 hours prior to receiving 4.5 mg Cd/kg IV. Within 24 hours of receiving the Cd, these authors found that the control rats (N = 6) had died, while all the ethanol pretreated rats (N = 7) had survived.

Follow-up studies revealed that ethanol pretreatment decreased the occurrence of Cd-induced serum transaminase activities by approximately 90%. Interestingly, the ethanol pretreated rats exhibited a 98% decrease in biliary excretion of Cd and a 31% increase in liver Cd levels. Studies on the ethanol-induced alteration in subcellular distribution of Cd indicated that the concentrations of Cd in nuclear mitochondrial and microsomal fractions were diminished by 60%, while there was a 90% increase in the cytosol levels. Ethanol also altered the cytosol distribution of Cd by decreasing the amount of Cd bound to high molecular weight protein (60 to 5%) and increasing Cd bound to metallothionein from 34 to 95%. Furthermore, the concentration of metallothionein in the liver was increased by 11-fold with the ethanol treatment. The authors concluded that "the mechanism of protection appears to be due to induction of metallothionein which sequestered Cd and decreases its accumulation in critical organelles and protein of the liver."

B. Cobalt

It has long been known that cobalt-fortified beer causes a special form of heart disease in chronic beer drinkers in the United States, Canada, and Belgium (Kesteloot et al., 1968). In a clinical sense this condition is characterized by a massive pericardial effusion, low cardiac output, and increased venous pressure, along with polycythemia in some instances. It should be noted that alcohol without cobalt fortification can cause heart disease, but these forms differ significantly from that caused by chronic consumption of the cobalt-fortified beer (Price, 1938; Brigden, 1957). Specific to the cobalt is the induction of polycythemia (Orten, 1935; Saikkonen, 1959) and production of massive pericardial effusion, as seen in animal models.

One of the major questions emerging from these initial observations was whether the alcohol enhanced the toxicity of cobalt or vice versa. To get a handle on this question, Derr et al. (1969) assessed the interaction between cobalt and ethanol on rat growth rate with particular emphasis on the heart. In this study, male albino rats (weight 60–70 g) were administered 10% ethanol in water, cobaltous chloride solution (1 mg Co/10 mL H_2O), or ethanol plus cobalt with exposures lasting 35 days. Their study revealed cobalt and ethanol interacted to cause a significant greater-than-additive effect in the depression of growth rate. In addition, the hearts of rats administered the ethanol–cobalt mixture were markedly smaller and contained higher concentrations of zinc.

The mechanism of the interaction was assessed in an analysis of the biochemical lesions induced by both agents. Derr et al. (1969) indicated that ethanol causes an increase in tissue NADH/NAD ratios (Cherrick and Leevy, 1965), while cobalt (as cobaltous salts) irreversibly complexes with the -SH group of dihydrolipoic acid and α-keto acid oxidase in vitro, thereby blocking the NAD-dependent reoxidation of dihydrolipoic acid to lipoic acid (Webb, 1968).

C. Hydrogen Sulfide

In 1979 Beck et al. decided to assess the capacity of ethanol to affect the occurrence of H_2S-induced CNS toxicity since both agents are known to adversely affect CNS function. Prior to this report of Beck et al. (1979), only one clinical occupational study had suggested that ethanol may enhance H_2S CNS toxicity (Poda, 1966). In the study of Beck et al., an 8% solution of ethanol in saline was administered intraperitoneally to male rats at dose levels of 0.33 and 0.66 g/kg body weight. When H_2S (800 ppm) was given 30 minutes following ethanol administration, the alcohol-exposed rats displayed an enhanced sensitivity as measured by time to unconsciousness. Both concentrations of ethanol reduced the time to unconsciousness by about 35–40%. The authors concluded that as a result of the widespread occurrence of H_2S exposures, the added risk caused by alcohol consumption should be clearly pointed out to those in danger of occupational exposure.

D. Lead

That ethanol consumption may affect lead toxicity was first suggested in 1966 by Cramer, who reported a higher incidence of lead poisoning among occupationally exposed (i.e., battery factory) workers with higher alcohol consumption when matched for time of employment and degree of industrial lead exposures. Lead-exposed Italian workers who were heavy drinkers were reported by Cardani and Farina (1972) to display a higher incidence of lead intoxication, as determined by changes in porphyrin metabolism. In this study it was not determined whether the workers were more susceptible to lead because of ethanol ingestion or whether they were exposed to greater amounts of lead via work practices or other means. In addition, there have been numerous cases in the rural southeastern United States of lead poisoning associated with ingestion of "moonshine" liquor (Morgan et al., 1966; Sandstead et al., 1970).

In an effort to establish a clear test of whether a causal relationship existed, Mahaffey et al. (1974) evaluated the effects of ethanol ingestion on lead toxicity in male albino Sprague-Dawley rats fed isocaloric diets with controlled nutritional content. In the 10-week study the ethanol (10%) and lead (200 µg Pb/mL) were administered in the drinking water with three sets of control animals (lead only, ethanol only, and unexposed). The results indicated that the joint exposure to ethanol and lead, as compared to lead alone, led to an increase in blood lead levels (47 vs 57 µg/100 mL), a decrease in hematocrit (46.0 vs 42.7%), increase in urinary ALA (33.0 vs 51.4 µg/24 hr), increase in renal Pb (6.42 vs 16.57 µg/gm wet tissue), noticeably more renal inclusions, increased liver Pb (1.136 vs 1.717 µg/gm tissue), and increased bone Pb (48.96 vs 56.20 µg/gm). Consequently the increase in renal lead levels was the most severely affected parameter. The mechanism by which ethanol ingestion caused an increase in tissue lead levels and the specific patterns of distribution was not evaluated.

E. Mercury

That ethanol consumption may affect the body burden and possible toxicity of mercury was first suggested as a result of a serendipitous discovery by Nielsen-Kudsk (1965) that the unscheduled ingestion of beer prior to mercury vapor inhalation of a volunteer subject reduced mercury retention. Consequently Nielsen-Kudsk (1965) conducted a controlled evaluation that established for the first time that ethanol ingestion resulted in a 50–70% reduction of the mercury vapor absorption by the same untreated subject.

In a subsequent study, Nielsen-Kudsk (1969) demonstrated that mercury vapor uptake by human blood was reduced to 30% of the control blood in the presence of 0.2% alcohol in the blood. This lack of uptake of mercury by the blood led Nielsen-Kudsk (1969) to hypothesize that the reduced lung absorption resulted from an alcohol inhibition of the conversion of elemental mercury to its ionic form by the red blood cells. Elemental mercury is present in the atmosphere as an uncharged monatomic gas. Following inhalation, it quickly crosses the alveolar membranes to enter the bloodstream, where it dissolves in plasma. As a result of its high diffusibility and high lipid solubility, the dissolved mercury vapor passes across the red cell or other cell membranes, where it becomes subject to oxidation to divalent ionic mercury. This form of mercury is able to react with numerous organic ligands.

This hypothesis was indirectly assessed in a 1973 report of Magos et al., who proposed that the alcohol effect might result from a decrease in the rate of oxidation, as noted above, or by a decreased transport rate from the alveolar lung spaces to the blood. Their study testing this alternative explanation involved the injection (IV) of rats with elemental mercury in solution. Rats thus exposed to mercury and pretreated with ethanol displayed an enhanced excretion of mercury in exhaled air, thereby supporting the original hypothesis of Nielsen-Kudsk (1965). A similar finding was reported by Dunn et al. (1978), who found that ethanol consumption enhanced the vapor excretion of ^{203}Hg by eightfold in CBA/J female mice.

These animal model studies were subsequently confirmed and extended in a controlled human exposure study in which volunteers ingested 1005 mL of beer (equivalent to 66 mL of ethanol) 30 minutes prior to exposure for 12–20 minutes to mercury vapor. The findings revealed that prior beer consumption

1. diminished mercury retention
2. enhanced rapid phase vapor loss by expiration
3. increased mercury storage in the liver
4. markedly diminished the extent of mercury uptake by the blood cells
5. eliminated the prompt storage of mercury by the blood (Hursh et al., 1980)

Follow-up experiments on mice and rats were again consistent with the human findings and revealed that alcohol exposure reduced mercury retention by 16–18%. At the conclusion of their paper, Hursh et al. (1980) provided a highly insightful and valuable 10-point series of general implications of the data:

1. Reduced retention due to ethanol pretreatment is probably caused by inhibition of the oxidation of Hg^0 in animal tissues, resulting in elevated concentrations of Hg^0 in venous blood returning to the pulmonary circulation.

2. Uptake of Hg^0 from inspired air from the blood depends on physical solution in the blood as it passes through the pulmonary circulation.

3. Dissolved vapor in blood is removed partly by oxidation to Hg^{++} in the red blood cells and partly by diffusion into, and oxidation by, other tissues.

4. Oxidation of dissolved Hg^0 in blood by the red blood cells makes a negligible contribution to uptake of vapor as the blood passes through the pulmonary circulation.

5. Deposition of mercury in lung tissue during exposure to vapor is by oxidation of vapor to Hg^{++} in lung tissue. This process is inhibited by ethanol, resulting in decreased lung deposition. Uptake of Hg^{++} from blood makes a negligible contribution in the lung mercury content.

6. Plasma concentration of Hg^{++} derives mainly from oxidation of Hg^0 in tissues other than red blood cells.

7. Ethanol causes an increase in Hg^{++} in liver tissue. Experiments in animals and liver homogenates indicate that the catalase activity in this tissue is so high that ethanol does not produce its characteristic inhibition. Consequently the increased availability of mercury vapor (due to lessened uptake by other organs and tissues) results in an increased Hg^{++} accumulation in the liver.

8. Increased exhalation of Hg^0 induced by ethanol two days after exposure is probably caused by the inhibition of the reoxidation of Hg^0 produced continuously by tissue reduction of Hg^{++}.

9. The simplest explanation for the effects of ethanol reported in these studies is that ethanol inhibits the oxidation of inhaled Hg^0 to Hg^{++} and likewise the reoxidation of Hg^0 produced in tissues by reduction of Hg^{++}.

10. The authors believe that, notwithstanding the clear evidence that alcohol reduces the uptake of mercury vapor, the control of the atmospheric concentration of mercury, rather than plying the workers with alcoholic beverages, remains the industrial hygiene method of choice. It may be that (as Dunn et al., 1978, suggest) the evaluation of inorganic mercury burden could be achieved by administering an oral dose of ethanol and measuring the exhaled mercury.

F. Nitrogen Dioxide

In 1983 a preliminary investigation by Early et al. assessed the effects of NO_2, ethanol, and their interaction on various aspects of cardiovascular function in male Sprague-Dawley rats. The NO_2 exposure consisted of 5 ppm for 72 hours; the ethanol was administered in the water at 10% for 13 days (of which 10 days were prior to NO_2 exposure). While the data are too preliminary to offer general conclusions, it appears that some interaction may have occurred with respect to atrial response to calcium.

III. ORGANICS

A. Acetaminophen

Acetaminophen (n-acetyl-p-aminophenol), which was synthesized in the later part of the nineteenth century, has become the principal alternative to aspirin in the United States and most other Western countries for the relief of minor aches and pains and febrile disorders (*Ann. Intern. Med.,* 1986). While considered safe for recommended use, acetaminophen was first recognized as a potential hepatotoxin in 1966 following consumption of 50 and 150 acetaminophen tablets in two suicide attempts (Davidson and Eastham, 1966). Since this initial observation, other reports have established that overdosing with acetaminophen is a risk factor in the development of hepatotoxicity (Volans, 1976; Meredith et al., 1981). In quantitative terms, the lowest doses of acetaminophen associated with hepatotoxicity appeared to be approximately 15 g, or about 50 tablets taken at a time (*Ann. Intern. Med.,* 1986).

While there seems to be a large margin of safety between recommended therapeutic doses and minimally hepatotoxic doses in humans, a series of case reports suggest that chronic alcohol consumption enhanced the risk of developing acetaminophen-induced liver damage (Baker et al., 1977; Goldfinger et al., 1978; Emby and Fraser, 1977; McLain et al., 1980; Kaysen et al., 1985; O'Dell et al., 1986; Lesser et al., 1986). More recent studies by Seef et al. (1986) have documented that some alcoholic individuals have developed severe liver damage with ingestion of 10 g of acetaminophen, and on occasion as little as 3 or 4 g, which begins to approach amounts ingested by some members of the general public. Initial complementary studies in adult male mice confirmed that a three-week prior exposure to ethanol in drinking water (10%) significantly increased the acute toxicity of acetaminophen (IP), with the LD_{50} decreasing from 876 mg/kg in controls to 518 mg/kg in the alcohol-treated mice. Histological analyses confirmed the occurrence of central hepatic necrosis typical for acetaminophen toxicity (McLain et al., 1980).

Since this initial study in an animal model, other investigators have repeatedly confirmed that chronic exposure to ethanol enhances the occurrence of acetaminophen-induced liver toxicity (Sato, C., et al., 1981a; Walker et al., 1983; Tredger et al., 1985). However, the situation became more complicated when it was determined that acute ethanol administration (in contrast to chronic exposure) actually was found to prevent acetaminophen-induced liver damage in rodents (Sato and Lieber, 1981; Sato, C., et al., 1981a; Tredger et al., 1985). These findings may be related to observations suggesting that chronic ethanol ingestion enhances the mixed-function oxidation of acetaminophen to reactive metabolites, while acute exposures to ethanol are known to inhibit drug metabolism in vivo and in vitro (Rubin et al., 1970; Sato and Lieber, 1981). Recently studies by Tredger et al. (1986) have begun to clarify and expand the metabolic picture of the relationship between chronic and acute ethanol exposures and acetaminophen toxicity, as illustrated in Figure 18.1.

The critical metabolic factor in the hepatotoxicity of acetaminophen seems therefore to be the formation of a bioactivated metabolite via the cytochrome P-450 microsomal mixed-function oxidase system and glutathione conjugation of the bioactivated metabolite. As noted above, chronic alcohol ingestion by animal models enhance the microsomal ethanol-oxidizing system. However, it is important to note that this increase in ethanol oxidation is associated with the occurrence of a distinct form of hepatic P-450. This dis-

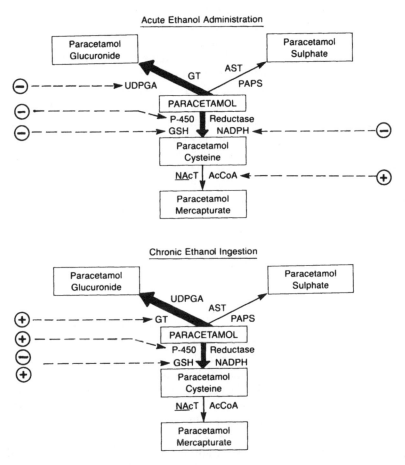

Figure 18.1. Recognized and proposed interaction of ethanol with paracetamol metabolism. ⊕, stimulation; ⊖, inhibition; AcCoA, acetyl coenzyme A; AST, aryl sulfotransferase; GSH, reduced glutathione; GT, UDP-glucuronyltransferase; *N*-AcT, *N*-acetyltransferase; PAPS, phosphoadenosylphosphosulfate; P-450, cytochrome P-450; reductase, NADPH-cytochrome P-450 reductase; UDPGA, UDP-glucuronic acid. From Tredger et al. (1986).

tinct enzymatic form of P-450 comprises about 5% of the total P-450 in the livers of untreated animals, but up to 20% in ethanol-treated animals. This distinct form of P-450—terminal P-450 3a in rabbits and P-450 j in rats (*Ann. Intern. Med.,* 1986)—also converts acetaminophen to a reactive metabolite that binds to glutathione (Morgan et al., 1983).

These collective findings have begun to piece together the complex web of ethanol–acetaminophen interactions. It is obvious that the elucidation of this interaction will have implications far beyond the ethanol–acetaminophen interaction to other interactions also of considerable public health significance. Bioactivation of nitrosamines to their carcinogenic form is now believed to proceed through the action of P-450 3a (Yang et al., 1985). However, it is uncertain at present whether it enhances the susceptibility of the chronic alcoholic to nitrosamine-induced cancer. Cytochrome P-450 3a has also been shown to be induced by acetone (Koop et al., 1985; Tu et al., 1983) and pyrazole (Tu et al., 1983). Chronic or acute exposures to these agents can therefore be expected to affect acetaminophen-induced toxicity.

Finally, the issue needs to be raised as to whether moderate consumption of ethanol on a daily basis enhances the risk of liver damage from ingestion of recommended doses of acetaminophen. A relevant study addressing this issue was published by Altomare et al. (1984), using the baboon as the animal model. Significantly, a dose of acetaminophen that maintained blood levels comparable to those recommended for humans did not cause hepatotoxic effects (i.e., morphological or functional) in baboons drinking alcohol (50% of total calories for 18 months). While chronic ethanol consumption altered acetaminophen metabolism consistent with metabolic changes reported by Tredger et al. (1986), the extent of exposure was apparently insufficient to cause detectable damage. Dose-response studies designed to detect threshold effects would be of considerable value.

B. Aflatoxin

In 1978 Glinsukon et al. assessed the capacity of an alcohol pretreatment to enhance the hepatotoxicity of aflatoxin B_1 in adult female rats. The rationale for this study was that "in some places, peanuts are taken as a snack with beer and whiskey. It is likely that the hepatotoxicity of aflatoxin B_1 contained in peanuts might be modified in persons who have been drinking alcohol."

In their study, Glinsukon et al. (1978) evaluated the effects of a range of ethanol pretreatments. More specifically, four oral doses of 1.0, 2.0, or 4.0 g/ kg ethanol were administered at 48, 42, 24, and 18 hours prior to i.p. administration of 1.0, 2.0, or 4.0 mg/kg aflatoxin B_1 dissolved in 0.5 mg/kg DMSO. The ethanol pretreatment significantly increased the aflatoxin B_1 induced hepatotoxicity at all doses of aflatoxin B_1 by two-to fourfold, as measured by plasma GPT and GOT. Histological evaluations also confirmed the enhanced effects of ethanol on aflatoxin B_1 induced hepatotoxicity. While the mechanism(s) responsible for the enhanced toxicity of aflatoxin B_1 by alcohol are unknown, the authors speculated that ethanol pretreatment may enhance the bioactivation of aflatoxin B_1.

C. Aspirin

1. GLYCINE CONJUGATION

In 1964 Childs and Lieberman reported that the urinary excretion of hippuric acid after oral or intravenous administration of sodium benzoate to human subjects was markedly diminished by ethanol. Six years later Amsel and Levy (1970) confirmed the original findings but also established that the ethanol (50 mL in humans) did not affect the renal clearance of hippurate but actually inhibits the formation of hippurate from benzoate. The mechanism of this inhibition was not biochemically established but was hypothesized as occurring through a depression of the activity of glycine N-acylase (which catalyzes the transfer of the drug to glycine) or a reduction of the availability of glycine to the conjugation system.

Since ethanol reduces the formation of glycine conjugation with benzoic acid, it may be hypothesized that ethanol may also have a comparable effect on glycine conjugation with salicylate since it is principally metabolized via glycine conjugation. In contrast to the interaction of ethanol and benzoic acid, there was no effect of ethanol on the maximum rate of formation of salicylurate from salicylate in human subjects given 50 mL ethanol. Amsel and Levy (1970) suggested that these findings are consistent with evidence indicating that the conjugation of benzoate and salicylate with glycine in humans act via different rate-limiting steps. More specifically, the maximum formation rate of hippurate from benzoate is limited by the availability of glycine (Amsel and Levy, 1969), while the maximum formation rate of salicylurate from salicylate in humans is believed to be controlled by the rate of activation of the substrate (Amsel and Levy, 1969; Tishler and Goldman, 1970).

2. GASTROINTESTINAL HEMORRHAGE

In an apparently unrelated direction of research to that noted above, epidemiologic findings suggest an interaction between aspirin and alcohol such that the combination of both result in a markedly enhanced risk of gastrointestinal hemorrhage (Needham et al., 1971; Goulston and Cooke, 1968; *Lancet,* 1970; Langman, 1970). According to Prescott and Critchley (1983a), the most significant interaction of ethanol and aspirin is the potentiation of the gastrointestinal toxicity of salicylates by ethanol. Both aspirin and ethanol cause gastritis and damage the gastric mucosal barrier. The interaction of these agents results in an enhanced damage of this gastric mucosal barrier, and this response has been deemed a synergistic one (Murray et al., 1974). Ethanol also potentiates aspirin-induced occult blood loss resulting, in extreme cases, in acute gastrointestinal hemorrhage (Goulston and Cooke, 1968; Needham et al., 1971). Furthermore, evidence exists linking ethanol as a precipitating factor in chronic gastric ulceration (Bock, 1976). Prescott and Critchley (1983a) claimed that "the importance of these aspirin/ethanol interactions is undoubtedly underestimated, and self-medication with aspirin-containing products for the prevention and relief of 'hangover' is commonplace."

D. Barbiturates

Although the first barbiturate, barbital (veronal), was reported in 1903 by Fisher and Mering (Goodman and Gilman, 1941), it was not until nearly 30 years later that Fuhner first reported on the combined action of ethanol and barbiturates. In this initial report, an instance was described where a person had consumed a glass of wine along with a tablet of Quinisal (chinium bisalicylosalicylium) and a tablet of propallylonal (Nostal). The individual fell into a deep coma but eventually recovered.

This initial finding lead to a series of reports that began to assess the nature of ethanol–barbiturate interactions in both animal models and human subjects. The first experimental study to follow up on the observations of Fuhner (1930) was that of Carriere et al. (1934) concerning the interaction of ethanol and phenobarbital. Using rabbits as experimental models, they reported that there was an antagonistic relationship between the two agents such that the ethanol treatment delayed or shortened phenobarbital-induced coma.

However, these initial and unexpected findings of Carriere et al. (1934) were not supported in a series of subsequent reports (Olszycka, 1935, 1936; Dille and Ahlquist, 1937; Jetter and McLean, 1943). For example, Allegri (1935) concluded that coma produced by phenobarbital or barbital sodium could not be antagonized by the administration of ethanol using rabbits, dogs, and guinea pigs. Furthermore, Olszycka (1935, 1936) established the occurrence of a marked potentiation of barbiturate (butethol—Neonal) induced sleeping time for both mice and rats. Olszycka developed a *potentiation coefficient,* which was defined by dividing the effect of the combined administration by the sum of the individual effects. The so-called potentiation coefficients achieved a value as high as 560. For example, one rodent group received an ethanol dose sufficient to produce a sleep of 2–3 minutes. Another rodent group received a dose of burethal to produce an 11-minute sleep time. However, the group receiving both treatments slept 191 minutes (i.e., potentiation coefficient of 14.5).

Dille and Ahlquist (1937) further confirmed the findings of Olszycka (1935, 1936). They found that ethanol administration (IV) markedly enhanced pentobarbital-induced (IV) duration of sleeping time in rabbits, and that the potentiation was greater with the use of smaller doses. In addition, the prolongation of the pentobarbital-induced sleeping time could not be explained by a prolonged half-life since the rate of pentobarbital elimination was unaffected by the presence of ethanol.

A continual series of studies followed that generally confirmed the basic observation that ethanol administration enhanced the pharmacologic and toxic effects of a variety of barbiturates via different routes of administration and in multiple experimental animal models (Ramsey and Haag, 1946; Smith and Loomis, 1951).

A series of publications soon followed that challenged this perspective. These studies, in fact, claimed that the interaction was not one of potentiation but of additivity. Thus, while the research of Fern and Hodges (1953) confirmed that ethanol enhanced the toxic and anaesthetic effects of amobarbital in mice, they claimed that the joint effects were simple additivity. Fern and Hodges (1953) claimed that

> most workers have reported the existence of [potentiation] between barbiturates and ethanol but they have not attempted to make any quantitative studies to determine the type of this [interaction]. In particular, experimenters who have reported a potentiation of the effects of barbiturates by ethanol have generally not used very convincing graphical or mathematical methods to substantiate their findings.

In fact, these authors claimed that the previous work of such researchers, including Olszycka (1935, 1936) and Jetter and McLean (1943), did not indicate potentiation, and that their own work, using the graphic mathematical approach of Gaddum, revealed clear additivity.

This perspective was soon supported by Gruber (1955), who critically evaluated the above cited investigations of the combined effects of ethanol and barbiturates and concluded that none of the previous studies provided conclusive evidence for the existence of potentiation (Eerola, 1961; Aston and Cullumbine, 1959; Graham, 1960). Criticisms were directed toward the use of the "coefficient of potentiation" of Olszycka (1936). For example, Aston and Cullumbine (1959) indicate that the method appears suspect since in studies by Olszycka (1936) with rats two doses of ethanol given individually and in combination provide a coefficient of potentiation of more than 9.0. Furthermore, Aston and Cullumbine (1959) challenged the use of the concept of potentiation "simply based upon the fact that sublethal doses of alcohol and a barbiturate, when administered in combination, will cause death in the treated animals (Jetter and McLean, 1943; Ramsey and Haag, 1946).

The issue of the nature of the interaction of ethanol with barbiturates is an important one to resolve. It is clear that, despite much debate over whether there is an additive or synergistic interaction, nearly every paper published has displayed a marked enhancement of the toxic or other endpoints measured, regardless of the species tested, specific barbiturates used, or highly diverse experimental methods employed, including variations in route of exposure, dosage, endpoint, duration of observation, etc. In general, it is clear than the LD_{50} of a variety of barbiturates decreases significantly as the animals are provided increasingly greater quantities of ethanol. Some studies reveal this relationship to be additive (amylobarbitone, secobarbitone, phenobarbitone) (Aston and Cullumbine, 1959; Fern and Hodges, 1953), while other reports (Wiberg et al., 1969) suggest greater-than-additive responses for such agents.

One frequently overlooked observation is that the magnitude of the interaction has been greater at low concentrations. While major challenges to the hypothesis that alcohol and barbiturates interact synergistically have been published (Fern and Hodges, 1953; Graham, 1960; Aston and Cullumbine, 1959), these papers have difficulty accounting for the findings of, for example, Wiberg et al. (1969), who showed that doses considerably less than the LD_0 for ethanol reduced the LD_{50} of multiple barbiturates by up to 50% (note that the estimated LD_0 was 5.0 mg/kg). Furthermore, while 3 g/kg of ethanol had no effect on sleeping time in rats, and 30 mg/kg thiopentone had a 24-minute average sleeping time, the two in combination caused an average sleeping time of 217 minutes. Even the findings of Eerola (1961), who claimed an additive relationship presented data that an LD_{o2} amount of ethanol along with an LD_{10} amount of hexobarbital produced a mortality rate of 70%.

Gruber (1955), a major supporter of the additivity perspective, displayed data indicating that subthreshold doses of ethanol markedly enhance the sleeping times of mice given Seconal. For example, a 50 mg/kg dose of Seconal induced, on average, about 11 minutes of sleeping time in male mice, while 1.95 mg/kg of ethanol had no effect. Yet the combined administration of both agents resulted in about 137 minutes of sleeping time.

Unfortunately, these striking findings suggesting greater-than-additive joint effects are often limited by the inadequate dose-response relationships, as in the case of barbiturate sleeping times prolonged by ethanol as reported by Wiberg et al. (1969), or lack of quantification, as in the case of Melville et al. (1966). Nonetheless, these studies were generally well designed and conducted.

The nature of the joint effects of ethanol and barbiturates, based on data in animal models, remains to some extent unresolved. In fact, this relationship may well be found to encompass both perspectives. For example, the relationship is known to be somewhat dependent on dose, and some barbiturates display a clearly stronger interaction than others. Thus it may be impossible to generalize about all barbiturates and experimental situations. One is reminded of the practical conclusion of Eerola (1961): "It has to be pointed out, however, that even if the synergism is of the additive type, the additive share of ethanol must not be forgotten when a person under the influence of ethanol is being given [barbiturate treatment]."

Despite the apparent conflict in the animal studies over additive vs synergistic responses, such studies have revealed a high degree of both qualitative and quantitative consistencies. Given that mice, rats, guinea pigs, rabbits, cats, and dogs have responded in a comparable manner to joint ethanol–barbiturate administrations, this strongly suggests that humans would also experience such an interaction. Indeed, as noted above, the very first study concerning alcohol–barbiturate interactions, which stimulated the spate of animal research that followed, was the report of Fuhner (1930) indicating that consumption of wine markedly enhanced the CNS-depressing effects of several barbiturates in a single individual. Since that time, clinical medicine has confirmed over and over again that joint exposure of humans to alcohol and barbiturates is potentially very deadly (Jetter and McLean, 1943; Fischer et al., 1948; Burrows, 1953). In quantitative terms, blood levels as low as 0.5 mg/100 mL of barbiturate combined with 0.1% alcohol have proven fatal (Gupta and Kofoed, 1966).

While much of the available information on the nature of the ethanol–barbiturate interaction is based in the clinical literature, a number of quasi-experimental studies exist (Elbel, 1938; Peter, 1939; Graham, 1960; Doenicke, 1962) that qualitatively confirm the animal data that ethanol consumption enhances the toxic and pharmacologic effects of a variety of barbiturates. Furthermore, Doenicke (1962) showed that as little as half a liter of beer was able to markedly enhance the occurrence of a decrement of psychomotor abilities induced by butabarbitone (200 mg).

1. POSSIBLE MECHANISMS OF THE ETHANOL–BARBITURATE INTERACTION

While no definitive answer exists concerning the precise biochemical nature of the interaction, several hypotheses have been proposed. For example, thiopentone is initially taken up by the brain and then released for storage in fat deposits; the duration of the thiopentone-induced sleeping time is directly associated with its level in the brain. Because ethanol prolonged the thiopentone sleeping time, Wiberg et al. (1969) speculated that it might slow down the distribution of thiopentone from the brain to the fat deposits.

In their studies Wiberg et al. (1969) found that ethanol enhanced the toxicity of barbitone more than four other barbiturates tested. Yet, according to Wiberg et al. (1969), it is unlikely that impairment of barbiturate metabolism is a significant factor in the interaction mechanism.

Another mechanistic consideration involves evidence that ethanol enhances the distribution of barbiturate into the brain. Limited research by Seidel (1967) indicated higher levels of pentobarbitone, but not of thiopentone or barbitone, in the brains of mice treated with ethanol, while Wiberg et al. (1969) found higher levels of phenobarbitone and barbitone in the brains of rats given the barbiturate and ethanol than in those given only the barbiturate. From a metabolic perspective, evidence exists that under some circumstances ethanol reduces the clearance of several barbiturates (e.g., pentobarbitone but not barbitone and thiopentone) (Seidel, 1967; Melville et al., 1966). However, research by Ramsey and Haag (1946) indicated that in mice barbitone did not alter the blood ethanol curves and that the blood barbitone curves were nearly identical both with and without ethanol.

2. CONCLUSION

That ethanol enhances the toxicity of a wide variety of barbiturates is now widely accepted. Based on a large animal model database and clinical experience, sufficient evidence exists to convincingly illustrate that this interaction occurs in people and can be deadly even at low levels of exposure. The nature of the interaction with respect to additivity and synergism remains to be resolved. However, it is quite possible that both types of interactions may be capable of occurring under different experimental conditions. The biochemical mechanism(s) explaining the interaction likewise remain to be elucidated, but evidence is emerging that suggests that multiple mechanisms, including those affecting tissue distribution, may be involved.

E. Benzene

That ethanol consumption may affect benzene hematotoxicity was first suggested by Synder et al. (1981), who speculated that since ethanol alters the activity of numerous enzymes, such as hepatic cytochrome P-450 and epoxide hydrase, and had been found to increase the mutagenic potency of dimethylnitrosamine (DMN) and benzo(a)pyrene (BP), it may also affect the hematotoxic potency of inhaled benzene (since benzene itself was thought to be bioactivated to a hematotoxic agent by the MFO). Consequently Synder et al. (1981) evaluated the extent to which ingestion of ethanol (5–15% in drinking water) affected the hematotoxicity of inhaled benzene (300 ppm, 6 hours/day, 5 days/week, for 18 weeks) in male C57B1/6J mice. The results indicated that peripheral red cells and lymphocyte cell counts decreased following benzene exposure, but that this effect was potentiated by the copresence of ethanol.

These findings were soon extended by the same research team (Baarson et al., 1982), using principally the same methodology but with the new endpoint of bone marrow and spleen cell effects. The results indicated that mice exposed to benzene had developed marrow and splenic cellular irregularities. The effects were even further aggravated in mice receiving both benzene and alcohol. Thus there was a more extensive decrease in lymphocytes in bone marrow and spleen and a decrease in granulocytes and normoblasts in the marrow, along with a greater degree of atypical cellular morphology in the alcohol–benzene exposed group. These findings of Baarson et al. (1982) suggest that the ingestion of ethanol enhances the hematotoxicity of inhaled benzene.

The mechanism by which ethanol enhances benzene hematotoxicity is believed to be based on its capacity to accelerate the hydroxylation of benzene and transformations of phenol into highly toxic metabolites (hydroquinone, catechol) (Nakajima et al., 1985). It is interesting to note that some other agents that induce cytochrome P-450, such as phenobarbital (PB) and 3-MCA, increase the rate of benzene metabolism but not its toxicity (Gill et al., 1979). This may be explained by observations that PB enhances the activity of enzymes involved in conjugation processes (e.g., UDP-glucuronyl-transferase), while ethanol has no effect on this conjugating enzyme (Sato and Nakajima, 1985). In addition, the liver microsomal benzene hydroxylase enzyme consists of at least two isozymes, with either a high or low K_m value. Ethanol preferentially enhances the low K_m form of the enzyme; PB preferentially enhances the high K_m form of the enzyme. Consequently ethanol will have a greater capacity than PB to form the toxic intermediate products at low levels of exposure to benzene.

F. Benzo(a)pyrene

Since chronic ethanol ingestion increases cytochrome P-450 content and microsomal enzyme activity in the liver and thereby affects the metabolism of numerous xenobiotics, Seitz et al. (1978) proposed to assess the effect of chronic ethanol ingestion on microsomal enzymes of the intestine. In their three-week study, male Sprague-Dawley rats were fed a liquid diet, with 36% of the total calories either as ethanol or as isocaloric carbohydrate. Following sacrifice, the investigators observed a threefold increase in intestinal microsomal cytochrome P-450 content and benzo(a)pyrene hydroxylase activity of the ethanol-fed rats. Furthermore, when benzo(a)pyrene was evaluated for mutagenic activity in the presence of microsomes from ethanol-fed rats, a clear dose-response increase in mutagenic activity occurred that was significantly greater than the responses of the pair-fed controls. The authors concluded their report by noting that the increased incidence of cancer among alcoholics may be partially related to the enhanced capacity of these individuals to bioactivate procarcinogens in the intestines.

G. Benzodiazepines

Since their introduction into clinical practice in 1960, the benzodiazepines (BZD), especially diazepam (DZP) (Valium) and chlordiazepoxide (CDP) (Librium), have become among the most prescribed drugs in the United States as well as in numerous other countries. The sheer number of people who use minor tranquilizers such as DZP is enormous. For example, it has been estimated that about 750,000 women of childbearing age in the United States are regular users of DZP or CDP (Chambers and Hunt, 1977). Estimates from Ontario suggest that 15% of women and 9% of men use a psychotropic drug at any given time, with diazepam accounting for about 53% of this use.

Given the prevalence of alcohol consumption in society (regular alcohol drinkers comprise about 40% of adults, with 15% being classified as heavy drinkers) and given the widespread use of both drugs (i.e., BZD and ethanol) within society and their depressant effects on the CNS, it is not surprising that considerable epidemiological and clinical research has been directed toward elucidating possible interactions of the BZDs and ethanol. Sellers et al. (1980a) have listed numerous studies that describe the interaction of single benzodiazepine and ethanol exposures administered concurrently (Tables 18.1 and 18.2). These studies have typically assessed how the combined effects of these two drugs affected human volunteers on various psychomotor tests such as reaction times, coordination, simulated driving tests, memory, and learning tests. In general, the collective findings have demonstrated that the joint drug exposure impaired performance. These findings are consistent with epidemiological studies that have linked traffic accidents with the copresence of alcohol and benzodiazepines (Sellers and Busto, 1982).

Evidence that alcohol and BZDs interact in humans have been repeatedly validated in animal model studies (Chan et al., 1982; Rajtar, 1977; Mehar et al., 1974; Giles et al., 1979; Edwards and Eckerman, 1979; Danechmand et al., 1967) often using short-term endpoints such as sleeping time. In the animal studies the effects of the joint exposures have been described as "additive or supra-additive." Follow-up metabolic studies have revealed that alcohol administration increases the plasma and brain levels of DZP by about sixfold in rats. In fact, Paul and Whitehouse (1977) reported that alcohol inhibited the metabolism of desmethyl diazepam, a pharmacologically active metabolite of DZP. It is believed that these observations may explain the supra-additive effect of the ethanol–DZP combination on motor coordination.

A similar study with ethanol and CDP revealed that joint exposure likewise resulted in increased blood and brain levels of CDP compared to mice injected with only CDP. However, in contrast to the previous report, the blood and brain levels of the N-demethyl metabolite of CDP (NDCDP) were not significantly different between the two groups. This led the investigators to conclude that the rise in CDP levels could only be minimally responsible for the supra-additive effects on ethanol sleeping time (Chan et al., 1979). However, the combined treatment caused a supra-additive decrease of cGMP levels in the cerebellum (Chan and Heubusch, 1982). This resulted in a prolonged period during which cerebellar cGMP levels were below 30% of normal, and this coincided with the increase in sleeping time, thereby suggesting a possible association of these factors. Of further significance is that this depression in cGMP levels noted above also occurred in mice injected with ethanol and NDCDP.

Efforts to uncover the molecular basis of the interaction of ethanol and BZDs have focused on receptor binding studies. Unfortunately, such efforts have not sufficiently clarified the nature of the interaction. It appears that ethanol exposure does affect BZD binding to receptors, but this effect varies according to dose and duration of exposure (Karobath et al., 1980; Kochman et al., 1981; Davis and Ticku, 1981; Ticku and Davis, 1981).

On the pharmacokinetic side it has been shown that acute ethanol administration increases the blood levels of orally administered clobazam, chlordiazepoxide, desmethyldiazepam (clorazepate), and diazepam; ethanol apparently accomplishes this increase by inhibition of the N-deethylation or hydroxylation but not glucuronide conjugation. However, the ethanol exposure may moderately affect pharmacokinetic parameters through small

Table 1. Pharmacodynamic Interactions of Single Benzodiazepine and Ethanol Doses Administered Concurrently

Drug (dose)[a]	Ethanol	Subjects	Tests	Results	References
			Psychomotor Tests		
Clobazam (20 mg)	38.2–39 g (calculated according to body weight)	8 men (mean age, 39.3 yrs)	Choice reaction, simple reaction, two-hand coordination	Combination increased impairment compared to ethanol alone but did not reach statistical significance	Taeuber et al., 1979
Diazepam (5 mg)	40 g	8 men (aged 18–25 yrs)	Digit symbol substitution	Significant* impairment when both drugs were taken in combination compared to either drug alone	Curry and Smith, 1979
(5 mg) (10 mg)	0.5 g/kg 0.8 g/kg	371 men 29 women (mean age ± SD = 22 ± 2.8 yrs)	Choice reaction, coordination	On their own the drugs did not affect performance, but in their various combinations they definitely impaired performance (coordination test I, significant*)	Linnoila and Mattila, 1973
(10 mg)	0.8 g/kg	20 men (aged 19–22 yrs)	Hand-to-eye coordination	Combination significantly* increased number of mistakes compared to either drug alone	Laisi et al., 1979
(10 mg)	0.78 mL/kg (96% ethanol)	8 men (aged 24–30 yrs)	Flicker fusion frequency, complex coordination, mirror tracing	Combination significantly* increased detrimental effects when compared to placebo; detriment was greater than with either drug alone	Morland et al., 1974
(10 mg IV)	0.7 g/kg + 8 hours of ethanol, 0.15 g/kg/hr	6 men (aged 21–32 yrs)	Pursuit rotor, letter cancellation, flicker fusion	Combination significantly* increased impairment as compared to diazepam alone at 2, 4, 6 and 8 hours for pursuit rotor and at 4 hours for letter cancellation	Sellers et al., 1980
(6 mg)	45 mL	12 men 6 women (aged 20–31 yrs)	Delayed auditory feedback, pursuit rotor	Combination significantly* impaired performance compared to diazepam alone in 2 of the 9 tests of delayed auditory feedback	Hughes et al., 1965

Table 1, continued

Drug (dose)[a]	Ethanol	Subjects	Tests	Results	References
(10 mg IV)	0.7 g/kg + 8 hours of ethanol, 0.15 g/kg/hr	6 men (aged 18–30 yrs)	Visual flicker fusion	Combination significantly** impaired performance compared to ethanol alone	Gander, 1979
Nordiazepam (5 mg) (10 mg)	0.17 g/kg	6 women (aged 21–31 yrs; mean 25.8 yrs)	Visuo-motor coordination	2 subjects showed enhanced performance and 4 subjects showed impaired performance	Nicholson, 1979
Temazepam (10 mg) (20 mg) (30 mg)	60 mg	10 men 8 women (aged 20–48 yrs; mean 32 yrs)	Choice reaction, digit symbol substitution	None of the three dose levels taken in combination with ethanol impaired performance	Hindmarch, 1978
Simulated Driving					
Diazepam (10 mg)	0.5 g/kg	70 professional drivers (aged 19–22 yrs)	Driving simulator	Combination significantly** increased number of collisions compared to either drug alone; combination increased number of drivers driving off the road	Linnoila and Hakkinen, 1974
(5 mg)	0.8 g/kg	36 subjects (aged 18–26 yrs; mean 20.6 yrs)	Uniwest driving simulator	Subjects with diazepam reacted significantly faster after ethanol administration than when sober	Milner and Landauer, 1973
Memory and Learning Tests					
Diazepam (10 mg IV)	0.7 g/kg + 8 hours of ethanol, 0.15 g/kg/hr	6 men	Word recall	Combination induced significant* impairment compared to diazepam alone	Sellers et al., 1980b

Source: Sellers et al. (1980a)
[a]IV = intravenous.
*$p < 0.05$
**$p < 0.01$

Table 2. Pharmacodynamic Interactions of Subacute Benzodiazepine Administration with Single Dose Ethanol

Drug (dose)[a]	Ethanol	Subjects	Tests	Results	References
			Psychomotor Tests		
Bromazepam (6 mg tid for 2 wks)	0.5 g/kg	19 men, 1 woman (aged 20–28 yrs; mean, 24.2 yrs)	Choice reaction, coordination, attention	Bromazepam impaired coordination in combination with ethanol	Saario, 1976
Chlordiazepoxide (10 mg tid for 2 wks)	0.5 g/kg	20 men (aged 20–23 yrs)	Choice reaction, coordination, attention	Combination impaired* coordination and attention but less than the diazepam–ethanol combination	Linnoila et al, 1975
Diazepam (5 mg) (10 mg for 2 wks)	0.5 g/kg	20 men (aged 20–23 yrs)	Choice reaction, coordination, attention	Combination markedly impaired performance even more than in acute experiments	Linnoila et al., 1974b
Flurazepam (30 mg for 2 wks)	0.5 g/kg	33 men, 7 women (aged 21–26 yrs)	Choice reaction, coordination, divided attention	Combination impaired coordination	Saario and Linnoila, 1976
Nitrazepam (10 mg for 2 wks)	0.5 g/kg	17 men, 3 women (aged 20–25 yrs)	Choice reaction, coordination, attention	Combination of the drugs was deleterious to psychomotor skills	Saario et al., 1975
(10 mg for 2 wks)	0.5 g/kg	6 men, 6 women (aged 21–38 yrs)	Kinetic visual activity, Stroop	No interaction could be clearly demonstrated	Roden et al., 1977
Methyloxazepam (20 mg for 2 wks)	0.5 g/kg	17 men, 23 women (aged 20–29 yrs; mean 22.9 yrs)	Choice reaction, coordination, divided attention, flicker fusion, proprioception, nystagmus	Combination significantly* impaired psychomotor skills	Palva and Linnoila, 1978
N-Desmethyldiazepam (5 mg for 2 wks)	0.5 g/kg	17 men, 23 women (aged 20–29 yrs; mean 22.9 yrs)	Choice reaction, coordination, divided attention, flicker fusion, proprioception, nystagmus	N-Desmethyldiazepam only occasionally enhanced ethanol-induced impairment of psychomotor skills	Palva and Linnoila, 1978

Table 2, continued

Drug (dose)[a]	Ethanol	Subjects	Tests	Results	References
Temazepam (10 mg) (30 mg for 2 wks)	60 mL	10 men, 8 women (aged 20–48 yrs; mean, 32 yrs)	Critical flicker fusion threshold, choice reaction, digit symbol substitution	Significant** depression of flicker fusion thresholds after 4 nights of repeated 30 mg of temazepam and ethanol; no significant changes with the 10-mg dose	Hindmarch, 1978
Memory and Learning Tests					
Bromazepam (6 mg tid for 2 wks)	0.5 g/kg	20 men (aged 20–25 yrs)	Immediate memory paired associated learning	Significant** impairment of learning	Liljequist et al., 1975
N-Desmethyldiazepam (10 mg tid for 2 wks)	0.8 g/kg	20 healthy volunteers (aged 20–25 yrs)	Learning acquisition, immediate recall of digit sequences	Significant** impairment of immediate memory and learning acquisition***	Liljequist et al., 1979
Oxazepam (15 mg for 2 wks)	0.8 g/kg	20 healthy volunteers (aged 20–25 yrs)	Learning acquisition, immediate recall of digit sequences	Significant** impairment of learning acquisition	Liljequist et al., 1979

Source: Sellers et al. (1980a).
[a]tid = 3 times a day
*p < 0.05
**p < 0.02
***p < 0.001

changes in drug absorption. Sellers and Busto (1982) have listed a number of kinetic interactions of acute ethanol exposures with single doses of BZDs (Greenblatt et al., 1978; Divoll and Greenblatt, 1981) and protein binding (Sellers and Busto, 1982).

While acute exposures to ethanol inhibit MFOs, chronic exposure to ethanol stimulates MFO activity. Given this situation, it is important to distinguish between studies with intoxicated chronic alcoholics and those of recently abstinent chronic alcoholics. Therefore, it follows that the capacity of chronic ethanol administration to induce MFOs is the best explanation of the reported decreases in chlordiazepoxide and diazepam blood levels (Sellers and Busto, 1982). Furthermore, these authors have concluded that the clinical consequences of an increased drug clearance is that patients will display a lower average steady-state drug level along with a reduced therapeutic or toxic effect. However, the clinical implications of benzodiazepam clearance is further complicated since a cross-tolerance exists between ethanol and benzodiazepam (Sellers and Kalant, 1978), thereby suggesting that chronic ethanol ingestion may not only diminish blood levels but decrease CNS sensitivity to benzodiazepam.

H. Caffeine

A number of studies have investigated the acute toxicity of joint exposures to caffeine and ethanol. Early studies (Pilcher, 1912; Cushny, 1924; Strongin and Winsor, 1935; Ritchie, 1965) suggested an antagonistic relationship. Follow-up studies by Alstott et al. (1973) using the principal metabolite of caffeine, that is, l-methylxanthine, confirmed the original impression that the effects were antagonistic in mice. When placed in human exposure terms, Alstott et al. (1973) stated that "in order to produce enough [l-methylxanthine] to effectively antagonize the effects of a dose of 1.0 g/kg of ethanol, a 70-kg man would need 1100 mg/kg of caffeine or approximately 187 cups of coffee, each containing 150 mg of caffeine."

I. Carbon Disulfide

Carbon disulfide (CS_2), widely employed as an industrial solvent, is known to be toxic to the nervous system, cardiovascular system, and liver (Beauchamp et al., 1983). Carbon disulfide is converted in the liver via the MFO system to carbonyl sulfide (COS) and an electrophilic sulfur metabolite (Dalvi et al., 1974, 1975). Acute toxicity studies by Chergelis and Neal (1980) have revealed that COS is at least 10-fold more toxic than CS_2, while the electrophilic sulfur metabolite is believed to be the causative agent in the occurrence of CS_2-induced hepatic necrosis (Dalvi et al., 1974, 1975).

Since bioactivation of CS_2 by the MFO system is an important factor in CS_2-induced toxicity and since alcohol consumption is known to induce the MFO system, it may be hypothesized that prior exposure to alcohol may enhance CS_2 toxicity. This idea was supported in 1984 by Dossing and Ranek, who suggested a synergistic interaction between CS_2 and aliphatic alcohols caused hepatotoxicity among workers in the chemical industry. Subsequent studies with rats indicated that ethanol potentiates the neurotoxicity of CS_2 (Opacka et al., 1985a, 1985b).

Synderwine et al. (1988) assessed the capacity of an 18-hour pretreatment by aliphatic alcohols (isobutanol, ethanol, methanol, and isopropanol) to affect pharmacokinetic parameters and toxicity of CS_2 in male Sprague-Dawley rats. The doses of alcohols selected were those known to significantly increase hepatic MFO activity. The data revealed that, with the exception of isobutanol, these alcohols significantly increased the level of CS_2 retained in the liver three hours after CS_2 administration. However, the methanol treatment significantly increased the level of CS_2 in the plasma, kidney, and brain. The magnitude of increases were not trivial, being in the range of 2.5- to 4.0-fold. The authors speculated that methanol may have enhanced CS_2 retention by increasing its solubility.

Such increased tissue levels are believed to serve as a reservoir for toxic routes of metabolism involving either dithiocarbamate formation or the MFO system. This conjecture was based on the knowledge that methanol is metabolized at about one-tenth the rate of ethanol (Lester and Benson, 1970) and that the dosage of methanol was higher than the other aliphatic alcohols. In addition, hepatotoxicity studies reveal that, of the aliphatic alcohols tested, only methanol significantly increased (twofold) plasma GPT levels. The increased plasma GPT in the methanol pretreated rats is associated with greater retention of CS_2 in the liver. Even though the methanol pretreatment enhanced the occurrence of CS_2-induced plasma GPT levels, alcohol induction of the MFO system by itself does not seem to be of sufficient magnitude to enhance CS_2 liver toxicity. Isopropanol metabolism studies by Rubin et al. (1984) suggested that the alcohol inducible isozyme of cytochrome P-450 appeared saturated at high doses, thus minimizing the formation of active metabolites at the elevated levels of exposure. Clearly further mechanistically oriented studies are needed in this area.

J. Carbon Monoxide

Deaths from fire-related activities frequently are associated with both the consumption of alcohol and smoking. Given that exposure to alcohol and the products of fire will occur in a large number of situations, what is known about the possible interactions of ethanol with some of the principal constituents of cigarette smoke, for example, carbon monoxide? Despite the relatively frequent occurrence of these joint exposures, this interaction has been little studied. However, according to Mallach and Roseler (1961), a distinct interaction occurs that is sufficient to cause death when mice are jointly exposed to high levels of ethanol and CO, such that ethanol consumption (i.e., blood level of 0.1–0.2%) significantly decreased the level of carboxyhemoglobin associated with death. Clearly more research is needed in this area.

K. Carbon Tetrachloride

1. ANIMAL MODELS

The hepatotoxicity of CCl_4 is contingent upon its bioactivation, most likely through the cleavage of the carbon chlorine bond in the endoplasmic reticulum and the production of free radicals (Slater, 1966). The toxicity of CCl_4 was found to be markedly altered by agents or treatments that affect the occurrence of enzyme induction within the endoplasmic reticulum. For example, treatment of rats with phenobarbital, a strong enzyme inducer, markedly enhanced the hepatotoxic effects of CCl_4 (Garner and McLean, 1969; Stenger and Johnson, 1971; Stenger et al., 1970). Likewise, it was shown that the protective influence of a low protein diet could be reversed by treating rats with DDT or phenobarbital (McLean and McLean 1966).

In addition to the marked potentiation of CCl_4-induced hepatotoxicity by enzyme inducers, CCl_4 hepatotoxicity can also be enhanced by pretreatment with aliphatic alcohols (Stewart et al., 1960; Klaassen and Plaa, 1966; Cornish and Adefuin, 1967; Traiger and Plaa, 1971; Strubelt et al., 1978a, 1978b). Marked differences have been reported between the potentiating capacities of the respective alcohols studied. Isopropanol, for example, enhanced CCl_4-induced liver toxicity to a much greater extent than ethanol, even though the dose of ethanol far exceeded that of isopropanol (Traiger and Plaa, 1971). However, despite these differences in potentiating capability, both alcohols appear to exert their maximum effect when administered 18 hours prior to CCl_4 exposure.

Subsequent studies revealed that, at least for ethanol, the parent compound was responsible for the potentiating effect. This was determined in studies using pyrazole, an inhibitor of alcohol dehydrogenase. When the pyrazole was administered 15 minutes before the ethanol treatment, the rate of elimination of ethanol was markedly reduced, while at the same time potentiating the effects of ethanol on CCl_4 hepatotoxicity when the ethanol was given 18 hours prior to CCl_4 treatment. Thus it was concluded that it was the unmetabolized ethanol that was involved in the potentiation (Plaa and Traiger, 1972).

When pyrazole was used with isopropanol, the elimination of the alcohol was again markedly reduced but the potentiating capability in contrast to ethanol, was markedly diminished. The increased isopropanol blood levels therefore could not account for the potentiation of the CCl_4. These results suggested that acetone, a principal metabolite of isopropanol, may be involved in the potentiation of CCl_4. Subsequent studies did in fact reveal that acetone could markedly potentiate the hepatotoxic effects of CCl_4 when given 18 hours before CCl_4 treatment and could account for much of the isopropanol induced potentiation.

The mechanism by which ethanol or acetone enhances CCl_4-induced toxicity remains to be elucidated but appears to be associated with the capacity of the liver microsomal enzymes to bioactivate the CCl_4. In fact, use of aminothiazole, which markedly depresses the drug-metabolizing enzymes of cytochrome P-450, significantly reduces the potentiating effects on CCl_4 hepatotoxicity (Plaa and Traiger, 1972).

Biochemical toxicological studies have revealed that acetone can affect the metabolism of various substrates. For example, acetone has exhibited an in vivo biphasic effect on aniline p-hydroxylation (Clark and Powis, 1974), delayed stimulation of DMN metabolism (Sipes et al., 1973a), and displayed in vitro inhibition of aminopyrine N-demethylation (Clark and Powis, 1974) and DMN N-demethylation (Sipes et al., 1973a). It has been further shown that ethanol pretreatment markedly enhanced the in vitro covalent binding of labeled CCl_4 (Maling et al., 1975), while isopropanol-enhanced cell toxicity is correlated with increased covalent binding in vivo of labeled CCl_4 to liver lipid and protein (Maling et al., 1974a).

Microsomes isolated from the liver of rats that had been pretreated with four doses of ethanol or a single dose of isopropanol or acetone displayed an enhanced capacity to bind labeled CCl_4 or $CHCl_3$ covalently (Sipes et al., 1973a). However, no changes were observed in the content of microsomal protein and cytochrome P-450 and NADPH-cytochrome reductase. These findings led Maling et al. (1975) to conclude that "all these pretreatments increase the formation of the carbon trichloro free radical by similar mechanisms." These researchers further substantiated this interpretation with findings indicating that CCl_4-induced diene conjugation of liver microsomal lipid increases following the ethanol pretreatment and is markedly increased after the isopropanol pretreatment or the pretreatment with pyrazole and a single dose of ethanol.

While these studies have not provided a fully elucidated understanding of how acetone potentiates CCl_4 hepatotoxicity, they do provide initial clues for further research to clarify such interrelationships.

2. STUDIES WITH HUMAN POPULATIONS

Studies with human populations have also indicated that individuals exposed to ethanol and CCl_4 are markedly more susceptible to the organ damaging-properties of CCl_4 (Von Oettingen, 1964). Furthermore, isopropanol potentiation of CCl_4 toxicity has also been shown to take place in humans. For instance, Folland et al. (1976) reported an industrial accident in which isopropanol-exposed employees suffered varying degrees of liver and kidney damage when CCl_4 was accidentally introduced into their work atmosphere. The isopropanol levels in samples of alveolar air from the workers was less than that found in the air of the plant, but the acetone concentration in alveolar air was high. These data, according to Plaa et al. (1982), indicated that the isopropanol was absorbed and converted to acetone in humans, as was predicted from the prior animal studies. Consequently Folland et al. (1976) concluded that the severity of hepatic and renal dysfunction in these workers was caused by an isopropanol-induced potentiation of CCl_4 toxicity.

Subsequent studies have confirmed and extended these initial, yet critical, discoveries. Of particular relevance has been the sustained efforts of Plaa and his associates. They have focused on attempting to determine the nature of the dose-response relationship in an effort to define better the nature of the potentiation threshold phenomenon (Plaa et al., 1982). In addition, these efforts have focused on validating the TLV for acetone when concurrent exposures to hepatotoxic haloalkanes such as CCl_4 are also present.

Pessayre et al. (1982) reported that rats treated i.p. with a mixture of nontoxic dosages of TCE and CCl_4 exhibited moderate-to-severe liver injury. They also found that the magnitude of the liver damage was dose-dependent for both agents. While a previous dose of TCE enhanced the hepatotoxicity of CCl_4, the administration of CCl_4 before the TCE did not increase the hepatotoxicity of a subsequent dose of TCE. According to Charbonneau et al. (1986b), the findings of Pessayre et al. (1982) were consistent with the notion that the hepatotoxicity seen following the administration of the TCE–CCl_4 mixture resulted from a TCE potentiation of CCl_4 hepatotoxicity. The TCE–CCl_4 interactive potentiation was then itself found to be potentiated by prior exposure to acetone using the rat model. Interestingly, acetone did not affect TCE hepatotoxicity.

The authors focused again on the industrial implications of these findings:

> [These findings] suggest that acetone TCE–CCl_4 interactions in the workplace need to be re-evaluated, particularly if the TCE and CCl_4 occur in concentrations that border those capable of producing hepatotoxicity. In inhalation studies performed with rats, the minimal effective concentration of acetone required for potentiation of CCl_4 toxicity appears to be about 2500 ppm (4-hour exposure) and is equivalent to an orally administered minimal effective dosage of 0.25 ml/ kg (Charbonneau et al., 1986a). This concentration is not far removed from the presently recommended TLV of 1000 ppm for acetone in the workplace. Furthermore, ongoing experiments in the laboratory suggest that the minimal effective dosage of acetone may be decreased as much as 5-fold when TCE–CCl_4 mixtures are employed instead of CCl_4 alone.

3. CLINICAL STUDIES

In 1921 Hall established the possible therapeutic use of CCl_4 in the treatment of hookworm, using dogs as the experimental model. Based on these very encouraging findings, his work was markedly extended both in animal model and human population studies. By 1923 Smillie and Pessoa noted that over 50,000 people had been treated with CCl_4 for hookworm control. It is interesting to quote their final conclusion:

> There is a wide variation in individual reaction to carbon tetrachloride. Large doses—10–20 cc—have been given to adults without producing apparent ill effects. Alcoholics are especially susceptible to the toxic action of the drug; so small a dose as 1.5 cc has produced severe toxic symptoms in an acute alcoholic.

4. CONCLUSIONS

There has been interest in whether ethanol may enhance the toxicity of a wide range of hepatotoxic agents. Limited studies suggest that agents that require bioactivation in the liver are likely to have their toxicity enhanced by ethanol pretreatment. This has been found to be the case for CCl_4, thiacetamide, and paracetamol (Strubelt et al., 1978a), and TCE (Cornish and Adefuin, 1967). Hepatotoxins that do not require bioactivation, such as phalloidin, α-amanitin, praseodymium, and galactosamine, do not have their toxicity significantly affected by ethanol pretreatment (Strubelt et al., 1978a). This apparent generalization, however, appears to be inconsistent with the lack of ability of ethanol pretreatment to enhance PCE and 1,1,1-trichloroethylene toxicity in rats (Cornish and Adefuin, 1967), even though it significantly stimulates their metabolism (Sato et al., 1980).

L. Chloral Hydrate

Chloral hydrate has long been known as an effective hypnotic (Maynert, 1965). In addition, chloral hydrate had been used in a 1:2.25 weight-to-weight combination with ethanol on a regular basis for women in labor at the Victoria Jubilee Hospital in Kingston, Jamaica (Parboosingh, 1960). However, there has been a long-standing debate over whether a potentiation between the hypnotic effects of chloral hydrate and ethanol exists (Shideman, 1954; Sollman, 1957; Adams, 1943).

In 1970 Gessner and Cabana conducted a comprehensive study of the interaction of the hypnotic and toxic effects of chloral hydrate and ethanol in male mice. Using the ED_{50} as the measure for hypnotic effects,[1] these researchers observed a significant potentiation of the hypnotic effect when these agents were present in a weight ratio that is equal to or greater than 1:7.2; simple additivity occurred with weight ratios smaller than 1:7.2.

From a metabolic perspective, chloral hydrate displayed a biological half-life in male mice of 12.0 minutes. Chloral hydrate is converted (50%) to trichloroethanol with a rate constant of 0.057 minute^{-1}. In contrast, trichloroethanol is metabolized considerably more slowly than chloral hydrate; its rate constant for disappearance is 0.0033 minute^{-1} and its biological half-life in mice is 211 minutes. Since the rate of trichloroethanol disappearance is one-tenth of its rate of formation from chloral hydrate, there is an in vivo accumulation of trichloroethanol following chloral hydrate administration. Co-administration of an equivalent amount of ethanol markedly increased the rate of chloral hydrate disappearance; the rate constant for disappearance increased from 0.057 to 0.075 minute^{-1} (30%). The rate constant for trichloroethanol formation was also increased by the co-administration of ethanol (up by 84%). Under such exposure conditions, this pathway accounted for greater than 70% of the administered chloral hydrate. It is the only pathway whose rate constant is increased by ethanol co-administration and, according to Gessner and Cabana (1970), more than accounts for the increase in the overall rate of chloral hydrate disappearance. Since the rate of trichloroethanol conjugation is unaffected by ethanol treatment, it naturally follows that greater accumulation of trichloroethanol will occur.

The above findings have important pharmacological and toxicological implications. Based on a comparison of ED_{50} values for righting reflex loss after IV administration, trichloroethanol is at least 1.18 times as potent a hypnotic as chloral hydrate. This may be due in large part to the greater lipid solubility of trichloroethanol (i.e., benzene–water partition coefficient of 3.5) compared to chloral hydrate (partition coefficient of 0.03). This greater lipid solubility would permit greater amounts of trichloroethanol to enter the brain. This enhanced degree of in vivo trichloroethanol accumulation, along with the higher hypnotic potency of trichloroethanol, is believed to adequately explain the potentiation by ethanol of the hypnotic effects of chloral hydrate.

M. Chloroform

Reports as far back as the early 1920s suggested that persons recovering from acute ingestion of ethanol appeared to display greater susceptibility to the liver- and kidney-damaging properties of the halogenated hydrocarbons than persons not exposed to ethanol (Lambert, 1922; Smillie and Pessoa, 1923; Lamson et al., 1924; Von Oettingen, 1955). However, because most of the studies assessing this interaction employed the simultaneous administration of both agents, an initial explanation proposed that ethanol facilitates the absorption of the hydrocarbon (Rosenthal, 1930; Stewart et al., 1960), thereby contributing to higher tissue levels of the toxic agents. It also appears that persons who consumed ethanol only several hours prior to (and not simultaneously with) their exposure to the toxic hydrocarbons also displayed enhanced susceptibility (Guild et al., 1958).

In an effort to validate and extend the temporal relationship suggested by Guild et al. (1958), but under the more rigorously controlled environment of an animal model study, Kutob and Plaa (1962) reported that pretreating (from 15 days to 12 hours) mice with intoxicating doses of ethanol, prior to challenging with a minimally hepatotoxic dose of chloroform, resulted in an increased incidence of abnormal liver function as indicated by a variety of endpoints (i.e., PB sleeping time, BSP retention, liver succinic dehydrogenase activity, and histological analyses). Of great interest is that the adverse effects occur at a time when there is no longer any detectable ethanol in the blood. Furthermore, the findings revealed that the ethanol pretreatment induced a time-related increase in liver lipids that resulted in a higher chloroform retention in the liver.

This enhanced concentration of chloroform in the liver was offered as the mechanism by which ethanol enhanced toxicity. These experiments by Kutob and Plaa (1962) offered a significant advance since they demonstrated an interaction between ethanol and chloroform (1) when simultaneous exposure was avoided, (2) when both agents were administered by different routes, and (3) using a lag time between exposures such that ethanol would be completely metabolized prior to chloroform exposure. These collective procedures established that the effect of ethanol on chloroform toxicity was not due to enhancing of chloroform absorption and were highly suggestive of a physiological response induced by the ethanol.

No enhancement of toxicity occurred if the chloroform was administered after the triglyceride levels returned to the normal or below normal range (i.e., 48 hours after ethanol exposure). Also, no enhancement of toxicity occurred if the dose of ethanol was reduced to a level (2–5 g/kg) that did not affect triglyceride levels.

This impressive series of findings by Kutob and Plaa (1962) was extended by Klaassen and Plaa (1966): an ethanol pretreatment of 12 hours was also found to enhance the occurrence of renal toxicity in male Swiss-Webster mice. It is important to note that chloroform caused renal effects at doses well below those altering liver function. Unfortunately, histological analyses did not specifically evaluate whether the ethanol pretreatment enhanced lipid accumulation. The enhancement of renal toxicity by ethanol was also found with respect to 1,1,2-trichloroethane, but not with CCl_4, PCE, TCE, dichloromethane, and 1,1,1-trichloroethane.

The initial studies from Plaa's laboratory were extended to a broader range of alcohols, with potentiation of chloroform with nephrotoxicity being seen with isobutanol, isoamyl alcohol (Watrous and Plaa, 1971), and isopropanol (Traiger and Plaa, 1974). However, in earlier research with CCl_4, it was found that its toxicity could also be enhanced by a variety of alcohols, with the most potent enhancing alcohol being isopropanol. The potentiating activity of isopropanol was subsequently shown to result from its metabolite, acetone. Consequently considerable interest was diverted toward whether acetone itself, and other ketones, would enhance the toxicity of chloroform. In fact, the capacity of a second alcohol, 2-butanol, to enhance halogenated hydrocarbon toxicity was attributed to the formation of the ketone, 2-butanone (Traiger and Bruckner, 1976; Traiger and Plaa, 1972a, 1974). These initial findings further substantiated the hypothesis that the administration or generation of abnormal quantities of ketonic substances markedly increases the sensitivity of experimental animals to haloalkanes.

This hypothesis was creatively supported by observations that alloxan-or streptozocitin-induced diabetes, starvation, or repeated ethanol administration enhanced the hepatotoxic effects of various haloalkanes (Hasumura et al., 1974; Hanasono et al., 1975a, 1975b; Maling et al., 1975) since each such treatment results in a significant increase in the systemic and/or hepatic content of ketonic bodies (Rerup, 1970; Lefevre et al., 1970; Bates et al., 1968; Lieber, 1974; Lawson et al., 1976). Further supporting evidence was that partial protection against liver damage by CCl_4 or chloroform was affected by insulin, thereby suggesting that the potentiation seen with alloxan was contingent on the metabolic sequelae (ketosis) of the diabetic state (Hanasono et al., 1975a, 1975b).

The ketone potentiation hypothesis has continued to be extended with other research. Hewitt et al. (1979) found that a pretreatment of kepone, a ketone, enhances chloroform toxicity, while mirex, a nonketone enzyme inducer, did not alter chloroform toxicity. Furthermore, N-hexane and its two principal metabolites, 2-hexanone and 2,5-hexanedione (HD), enhanced hepato- and nephrotoxicity in male Sprague-Dawley rats (Hewitt et al., 1980b). 2,5-HD was also found to enhance the toxicity of chloroform as well as other chlorinated hydrocarbons (i.e., CCl_4 and 1,1,TCEA) in primary rat hepatocyte culture (Jernigan et al., 1983). Sex differences were found in the capacity of 2,5-HD pretreatment to potentiate various chlorinated hydrocarbons. For example, 2,5-HD pretreatment potentiated CCl_4 and deuterated CCl_3 induced hepatotoxicity in male mice, but not with TCE, 1,1,2-TCEA, and PCE. In contrast, using female CR mice, such treatments resulted in potentiation of chloroform, deuterated CCl_3, CCl_4, TCE, and 1,1,2-TCEA, but not PCE. This apparent striking sex difference in mice is of considerable interest since sex differences in mice are not very common—especially in comparison with rats (Jernigan and Harbison, 1982).

Given the consistent observations that ketonic agents may enhance chloroform toxicity, systemic studies have revealed that the potential of ketones to enhance the severity of chloroform-induced toxicity is positively associated with the length of the carbon chain (Hewitt, W.R. et al., 1983a). However, the mechanism by which such agents enhance chloroform toxicity has been addressed to a limited extent. Earlier research by Branchflower and Pohl (1981) presented evidence that MBK may enhance chloroform toxicity by reducing hepatic GSH levels and enhancing the formation of $COCl_2$, a bioactivated metabolite. Furthermore, acetone is known to induce various hepatic MFOs, which may increase covalent binding of the $COCl_2$ to liver microsomal protein (Sipes et al., 1973a). In addition, Kramer et al. (1974) have shown that continuous exposure of mice for 2–4 days to N-hexane increased MFO.

Besides ketonic agents enhancing chloroform toxicity, a variety of enzyme inducers have also been reported to potentiate the occurrence of chloroform-induced adverse effects. These include phenobarbital, 3,4-BP, 3MCA, PCB, PBB, and DDT (Scholler, 1970; Brown, 1972; Brown et al., 1974; Illet et al., 1973; Lavigne and Marchand, 1974; Kluwe and Hook, 1977, 1978a; Gopinath and Ford, 1975). Presumably these agents enhance chloroform induced toxicity by increasing its rate of activation by the MFO.

Enhancement of chloroform-induced toxicity in rats by diethylmaleate is mediated by a reduction in hepatic GSH content (Brown et al., 1974).

In contrast to the enhancement of the hepatic toxicity of chloroform by the above enzyme inducers, MCA, dioxin, and PCBs actually reduced the hepatotoxic effects of chloroform, while PB pretreatment had no effect. PBBs, on the other hand, enhanced the susceptibility of mice to the nephrotoxic effects of chloroform, CCl_4, TCE, and 1,1,2-TCEA (Kluwe and Hook, 1978b; Kluwe et al., 1978a, 1978b).

Questions and remaining perspectives based on these collective studies include the following:

1. Initial studies by Kutob and Plaa (1962) clearly revealed that ethanol pretreatment resulted in the build-up of liver lipids, and that this was associated with markedly higher hepatic chloroform levels. This was believed to have contributed to the susceptibility of the liver to chloroform toxicity. However, this observation has never been validated and extended since the original findings. It appears that once ethanol was found to enhance the covalent binding of chloroform to liver microsomes that resolution of the mechanism of toxicity focused on this and related research approaches. However, the two observations are not at variance with each other and may provide complementary perspectives.

2. Many of the studies using chloroform involved the use of corn oil as a vehicle. Since corn oil has been implicated as a factor promoting chloroform-induced liver cancer in mice but without apparent effect on chloroform-induced renal cancers in rats, it would be of value to assess how corn oil as a vehicle affects the capacity of ethanol to enhance chloroform-induced liver and kidney cancer.

3. Starvation enhances chloroform toxicity because of the formation of ketone bodies. Thus the question arises as to whether people on extreme weight-reducing diets may be at enhanced risk to chloroform and related solvents. This could be efficiently tested in an experimental setting.

4. The issue is also raised about the presence of ketones in natural food products and whether such levels are sufficiently high to enhance the toxicity of halogenated solvents.

5. While most of the attention has focused on the effects of ethanol and ketones on chloroform toxicity, a 1982 report by Harris et al. revealed that the joint administration of chloroform and CCl_4 resulted in a synergistic occurrence of liver toxicity, as measured by SGPT and ethane expiration and triglyceride values. A similar synergistic interaction was seen with bromoform and CCl_4. According to the authors, these findings may be particularly significant since they show that subthreshold doses of chloroform and CCl_4, when administered together, may cause liver damage. If such is the case for a chloroform–CCl_4 interaction, it would be of great value to now superimpose the enhancing influence of ethanol and ketonic bodies.

6. The relevance of these studies for humans in terms of dosage is important to consider. According to Sato et al. (1980), a typical ethanol experimental dose of 4 g/kg, which stimulates liver enzymes of rats very effectively and approaches the 5 g/kg initially used by Kotub and Plaa (1962) and Klaassen

and Plaa (1966), is equivalent to drinking a whole bottle of spirits. This situation is not commonly encountered but is by no means a rare occurrence for some humans. Although one must be cautious to extrapolate findings from rats to humans, it would not be unreasonable to assume that excessive ethanol intake may also alter the susceptibility of humans to toxic effects of chemical substances through affecting their metabolism. Nevertheless, Hewitt and Brown (1984) contend that the occurrence of an acute nephro- and/or hepatotoxic interaction between a ketonic solvent and a halogenated hydrocarbon seems unlikely at the threshold limits currently in effect. However, they implied that renal injury may be possible as a result of chronic low-level exposure, and that interactions of halogenated hydrocarbons and ketonic solvents need to be considered when assessing the hazard of an occupational setting.

7. The extrapolation issue for the hepatotoxicity of chloroform as enhanced by ketonic solvents maybe become a highly debated area since ketones such as 2-hexanone are capable of themselves producing renal tubular degradation at high doses. The lesions produced by these ketonic agents are similar to those reported in renal tissue of rats following inhalation of petroleum hydrocarbons (Carpenter et al., 1975, 1977) or co-administration of JP-5 jet fuel (Parker et al., 1981). In fact, Vernot and Pollard (1983) have isolated 2-hexanone and various odd carbon number methyl ketones from the urine of Fisher 344 rats exposed to JP-4 jet fuel. Of extrapolative relevance is that the lesions produced in the kidney by ketonic agents such as 2-hexanone have been found in male rats (Brown and Hewitt, 1984; Hewitt et al., 1980b; Hewitt and Brown, 1984) without any other model used. The renal lesions produced by jet fuel in the kidney of rodents leading to renal cancer is principally a phenomenon of the male rat kidney and not found in the female rat or male and female mice. Such sex- and species-specific responses and the lack of occupational epidemiologic validations have led to widespread doubt over the public health relevance of petroleum-related renal tumors. It would appear that such a question may be raised in the case of ketonic enhancement of hydrocarbon-induced renal toxicity until it is clarified by studies in female rats and male and female mice.

8. The capacity of ketonic solvents to enhance the toxicity of other nephrotoxic agents remains to be more comprehensively addressed. For example, acetone is known to enhance the hepatotoxic effects of 1,1-dichloroethylene, bromodichloromethane, and dibromochloromethane in rats, but without notable increase in nephrotoxic effects (Hewitt and Plaa, 1983; Hewitt, W.R. et al., 1983b). No enhancement of renal toxicity by ketonic solvents was reported with hexachloro-1,3-butadiene (HCBD) or $K_2Cr_2O_7$ (Hewitt and Brown, 1984). However, whether this lack of enhancement was a true lack of interaction or a failure to apply an effective dosage regimen is unknown. The latter alternative was suggested since MacDonald et al. (1982) and Hewitt and Plaa (1983) demonstrated that the enhancement of 1,1,2-trichloroethane and 1,1-dichloroethylene hepatotoxicity by acetone follows a biphasic dose-response curve with low levels enhancing, and high levels protecting against, hepatotoxicity.

9. According to Hewitt and Brown (1984), chloroform-induced hepatotoxicity apparently results from its bioactivation to the reactive intermediate, phos-

phene, which causes a depletion of GSH, leading to the formation of covalent adducts with hepatic macromolecules (Mansuy et al., 1977; Pohl et al., 1977, 1980). With respect to chloroform-induced kidney damage, it is known that the mouse kidney bioactivates chloroform to a nephrotoxic intermediate, which produces proximal tubular damage via a mechanism similar to that delineated in the liver (Smith and Hook, 1983). However, while ketonic solvents increase the renal cortical bioactivation of chloroform in renal tissue, there is uncertainty over what the mechanism of toxicity is, since the actual binding in the kidney for 2-hexanone pretreated rats is about 10% that seen in the liver of 2-hexanone pretreated rats, even though both tissues are comparably damaged. Nonetheless, it is quite possible that the kidney is more sensitive than the liver (Hewitt and Brown, 1984).

N. Cocaine

The pharmacological effects of cocaine have been well characterized as a powerful and rapid stimulation of the central nervous and cardiovascular systems. Excessive doses of cocaine may result in convulsions and unconsciousness, followed by death due to respiratory depression. In addition to acutely toxic effects, a cocaine-induced latent toxicity has been reported in phenobarbital-pretreated mice. This phenomenon is characterized by a decrease in acute (3 hours) lethality of cocaine and the occurrence of a latent (1–7 days) lethality (Evans and Harbison, 1978). According to Smith et al. (1981), the increase in latent toxicity was caused by a massive hepatic periportal necrosis, which was both dose- and time-dependent and associated with MFO activity. In fact, when the MFO activity was inhibited by SKF-525A, the latent toxicity of cocaine was blocked in phenobarbital-pretreated rats.

Since the cocaine latent toxicity was enhanced by enzyme inducers such as phenobarbital, 3 MCA, and PBBs, Smith et al. (1981) hypothesized that ethanol, another inducer of MFO activity, could also affect cocaine-induced latent toxicity. In their study Smith et al. (1981) administered ethanol (4.3% in a liquid diet for 5 days) to adult male mice. This was sufficient to significantly increase hepatic cytochrome P-450 levels without causing liver damage. The ethanol treatment significantly diminished cocaine-induced acute lethality from 67 to 23%, while potentiating latent (1–7 days) cocaine-induced hepatotoxicity.

Smith et al. (1981) explained their findings by noting that cocaine hepatotoxicity is contingent on the activity of the MFO. Inducers of MFO and inhibitors of esterase result in greater proportions of drug being metabolized by the MFO system, thus leading to hepatotoxicity. More specifically, cocaine is apparently converted to norcaine, an N-demethylated metabolite, which is a more potent hepatotoxin than cocaine and does not require bioactivation to cause toxicity. It is interesting to note that the location of the lesion after alcohol pretreatment is in the centrilobular area, whereas phenobarbital induction or chronic cocaine treatment results in periportal damage.

O. Dimethylhydrazine

In 1928 Krebs reported that the direct instillation of 50% ethanol into the rectum of mice caused rectal adenocarcinomas similar to those found in humans. However, exposure to ethanol in the drinking water (20%) did not result in the development of such tumors following 15 months of examination (Ketcham et al., 1963). Nevertheless, a number of epidemiological investigations have established an association between beer consumption and rectal cancer (Enstrom, 1977; McMichael et al., 1979; Pollack et al., 1984; *Lancet,* 1980). The proposed biological mechanism involves the stimulation of bile salt excretion by ethanol. Since bile salts are cocarcinogens, the beer may be acting in an indirect way to enhance the capacity of bile salts to interact with known colonic carcinogens.

This theory was tested by Nelson and Samelson (1985) using the model colonic carcinogen 1,2-dimethylhydrazine (DMH) in the Sprague-Dawley rat model. More specifically, the researchers conducted two separate studies in which the effect of ethanol on experimental colonic cancer was assessed. This was followed by a study on the effect of commercially available beer on colon cancer. In one experiment the ethanol was provided in the drinking water (5%), while the DMH was given in 16 weekly SC injections of 15 mg/kg. The rats were sacrificed 22 weeks after the first DMH injection (or about 30 weeks of age). The beer study was similar in design, with the only change being 10 weekly injections of 20 mg/kg DMH and sacrifice 14 weeks after the last injection (or 22 weeks of age).

The findings revealed that the ethanol did not cause any statistically significant changes in the number of colonic tumors. More specifically, 77 colonic tumors were found in the DMH-alone group versus 86 colonic tumors in the DMH plus ethanol group (P = 0.7). In contrast, in the beer study, the rats given the beer treatment developed significantly fewer gastrointestinal tumors (100 vs 67, or 2.9 vs 1.3 tumors/rat; P < 0.03), and fewer colon tumors/rat (2.33 vs 1.25).

Nelson and Samelson (1985) concluded that one should be reluctant to describe beer as a major risk factor in rectal cancer given the reduction in the number of colonic tumors induced, the absence of rectal tumors, and the marked decrease in the total number of gastrointestinal tumors. In fact, their findings revealed that the beer was actually protective. The relevance of these studies, nonetheless, requires further clarification since the experimental model, while of great interest, is of questionable utility to human populations, given the extremely high doses of DMH used.

P. Dimethyl Sulfoxide

Use of dimethyl sulfoxide (DMSO) in the general population is widespread. One of the initial subjective observations of an individual receiving DMSO treatment is the garliclike taste soon after treatment. It was this very observation that led to the hypothesis that DMSO and ethanol interact within the human body. More specifically, while most people experience the garliclike taste after DMSO administration, this taste sensation disappears after ingesting small doses of alcohol. This rather unsuspected finding was substantiated in subsequent studies, which established that alcohol consumption markedly reduced the expiration of the metabolite dimethyl sulfide (Jentschura, 1965). Such findings have led to the suggestion that ethanol inhibits the formation of this metabolite.

Other research has also indicated that DMSO and ethanol interact in ways that affect the toxicity of each other. A 1965 report by Boost revealed that patients who had received DMSO developed symptoms of intolerance after only moderate ingestion of alcoholic drinks. In 1967 Mallach revealed that prior consumption of ethanol (1 hour) markedly enhanced the acute toxicity of DMSO in white mice, while simultaneous administration of both agents did not. The mechanism of this time-related interaction of DMSO and ethanol has been hypothesized to involve the influence of a metabolite of DMSO on the formation of biogenic amines and the disturbance of amino acid metabolism.

The effect of DMSO on alcohol also deserves some mention. While DMSO does not affect the absorption of ethanol from the gastrointestinal tract, research with human subjects indicates that DMSO enhances the elimination of ethanol from the body. Even though the DMSO enhances the elimination of alcohol (Mallach, 1967), it paradoxically enhances alcohol-induced psychomotor impairment (Heck et al., 1966; Mallach, 1967). In addition, DMSO also appears to remove the odor of alcohol from respiratory air, although the ethanol is still detected in the air by means of chemical analysis (Mallach, 1967).

Thus, it appears that DMSO and ethanol interact in numerous ways in several species, affecting multiple endpoints. Given the widespread use of these two agents within society and these initial efforts indicating metabolic and toxic interactions, further research in this area would appear justified.

Q. Dinitrotoluene

In 1942 McGee et al. assessed the relationship between ethanol consumption and occupational toxicity associated with dinitrotoluene (DNT). This was stimulated in large part because of the increase in symptoms of chemical intoxication following the introduction of large numbers of new workers into the manufacturing of DNT and the prevailing observation that it was a common experience in World War I that habitual users of alcoholic beverages did not tolerate exposure to toxic chemical compounds.

Of the 154 men surveyed, 23 noted a reduced tolerance for alcohol and 31 reported that their DNT toxicity symptoms were aggravated by drinking alcohol. In fact, some workers indicated that they were unable to drink any alcoholic beverages during the initial 2–3 hours after the end of the work shift without a strong reaction. In contrast, 83 workers did not sense any interaction between their alcohol consumption and job-related DNT exposures and effects. The final 17 workers were nondrinkers.

These initial clinical observations suggest the need for better animal model research and rigorous epidemiological investigations in order to properly evaluate the proposed association between alcohol and DNT.

R. Ethylene Glycol and Related Agents

Ethylene glycol (antifreeze) is a potent renal toxin in a wide variety of animal models and humans. The renal toxicity is mediated through the formation of the metabolite oxalic acid, which precipitates within the renal tubules and causes toxicity. Ethylene glycol has a long history of abuse in the United States; numerous people have died from its ingestion. It is interesting to note that ethanol protects certain animal models from some of the acutely toxic effects of ethylene glycol (Peterson et al., 1963). In fact, prompt "clinical" treatment with ethanol of animal models acutely and lethally intoxicated with ethylene glycol often results in their survival. Several years following the original publication of Peterson et al. (1963), the capability of ethanol treatment to prevent acute ethylene glycol intoxication in humans was confirmed (Wacker et al., 1965).

Since the acute toxicity of ethylene glycol as noted above appeared to be related to the metabolic production of oxalic acid, the speculation arose that ethanol exposure may prevent the formation of oxalic acid. Consequently Blair and Vallee (1966) demonstrated that the oxidation of ethylene glycol is competitively inhibited by ethanol. In fact, in animals protected from ethylene glycol by ethanol treatment, most of the ethylene glycol is excreted unchanged in the urine (Peterson et al., 1968).

Based on the original findings of an ethanol–ethylene glycol interaction, Peterson et al. (1968) evaluated the capacity of ethanol to protect against acutely toxic effects of monofluoroethanol and monochloroethanol in Sprague-Dawley rats and squirrel monkeys. As in the case with ethylene glycol, the ethanol again was found to be protective. More specifically, in rats treated with ethanol (2 mL/kg i.p., 15 minutes after receiving the halogenated ethanol), the LD_{50} for fluoroethanol was increased 20-fold, and for chloroethanol 4-fold, over the control group. Similar protective action against the lethal effects of these agents was also found in monkeys.

The mechanism of the protection of ethanol again seemed to be related to the observation that both of the halogenated ethanols are substrates of liver alcohol dehydrogenase (Bartlett, 1952; Bernheim and Handler, 1941). It is likely that the ethanol competitively inhibits the metabolism of these agents, as in the case of ethylene glycol. Consequently it is likely that the parent compound may be excreted without formation of the more toxic metabolite. While the reason for the greater protection against fluoroethanol is unknown, it may be related to a greater inherent toxicity of the chloroethanol as a parent compound.

S. Isoniazid

In 1981 Chin et al. set forth to evaluate the hypothesis that concurrent ingestion of ethanol enhances the occurrence of acute isoniazid (INH) intoxication in mongrel dogs. This study was prompted by clinical observations that ethanol potentiates the effects of INH overdose and enhances lethality. In this study 4 of the 6 control animals challenged with 50 mg/kg of INH had repeated convulsions, and 2 died. The convulsions occurred after a mean latency of 98 minutes and were clonic-tonic in nature. In contrast, pretreatment of 1 hour IV with ethanol (2 g/kg) decreased the severity of seizures and altered the seizure pattern to clonic convulsions but did not significantly increase the survival rate. At the higher INH dosage of 75 mg/kg, the ethanol (2 g/kg) pretreatment again modified the seizure pattern to clonic convulsions but did not affect significantly the mortality rate.

Thus the findings of the Chin et al. (1981) study demonstrated that ethyl alcohol not only didn't enhance the acute toxicity of isoniazid but also provided an anticonvulsant effect to INH intoxication, as seen in an increased trend for latency of convulsions and decreased severity of INH-induced convulsions. Consequently these data with the dog model "contradict the clinical impression that acute INH toxicity may be enhanced in alcohol intoxicated persons." In partial contrast to the findings of protective effects of an ethanol pretreatment in the dog model, ethanol (1.5 g/kg) was ineffective as an antidote when ingested 30 minutes after a 75 mg/kg dose.

T. Menadione

In 1988 Ganey and Thurman assessed the capacity of ethanol to enhance the hepatotoxicity of menadione in perfused rat liver. The basis of this research is found in the observation that menadione is an oxygen-dependent liver toxin that causes biochemical lesions of periportal regions of perfused liver, possibly because of redox cycling. Since redox cycling requires reducing equivalent and ethanol metabolism generates reducing equivalents, a study of the interaction of ethanol and menadione was justified. In their study it was shown that ethanol treatment enhanced the occurrence of redox cycling and hepatotoxicity due to menadione, as measured by increasing pyridiol nucleotide fluorescence and maximal increases in oxygen and LDH values.

U. Methomyl

Long-term administration of ethanol has been reported by Smyth et al. (1967) to cause a decrease in brain cholinesterase activity. Given this circumstance it would be of some interest to assess whether ethanol consumption may interact with agents known to decrease brain cholinesterase activity. Consequently in 1979 Antal et al. reported on a study in which the effects of ethanol on methomyl, a carbamate insecticide, were assessed. Of particular interest were

1. the confirming observations that ethanol treatment (10% aqueous solution) significantly reduced (20–25%) brain acetylcholinesterase activity in male and female rats
2. the observation that while methomyl significantly reduced brain acetylcholinesterase activity in both sexes, only the male rats displayed a further marked decrease in enzyme activity after the combination of the methomyl and ethanol (no mechanistic explanations were offered to account for this striking sex difference)

V. Morphine

Ethanol pretreatment increases the concentration of free or unmetabolized morphine in the liver of rats given a single dose of morphine sulfate. Research by Steup and Forney (1988) with adult male Sprague-Dawley rats suggests that such observations are the result of the inhibition of liver metabolism of morphine rather than an enhanced delivery to the liver of morphine. (See bottom of page 510 concerning additional morphine/ethanol interactions.)

W. Nitrosamines

There has been an enormous interest in the role of ethanol ingestion on the occurrence of human cancer. Despite the limitations of epidemiological data, it is widely accepted that alcohol consumption can lead to cancer of the liver, esophagus, oropharynx, and larynx. The magnitude of the role of alcohol consumption in human cancer is believed to be of serious consequence, accounting for 7% of all cancer deaths in men and 3% in women for those sites related to alcohol consumption (Driver and Swann, 1987).

The mechanism by which alcohol consumption affects the occurrence of cancer has been extensively studied. It appears able to affect the process of carcinogenesis at different stages of initiation and promotion, as illustrated in the simplified scheme outlined in Figure 18.2. While the role of ethanol in the process of carcinogenesis remains to be clarified, Seitz (1985) has argued that ethanol may act to enhance the potency of various chemical carcinogens. The interaction of alcohol with chemical carcinogens has been studied most extensively with respect to the induction of hepatic carcinogenesis using nitrosamines in rodents. These collective studies have revealed that when nitrosamines were used to induce hepatic tumors, the occurrence of hepatic carcinogenesis often seemed to be not affected by ethanol ingestion, and at times some inhibition of carcinogenesis was seen. In contrast, the ethanol

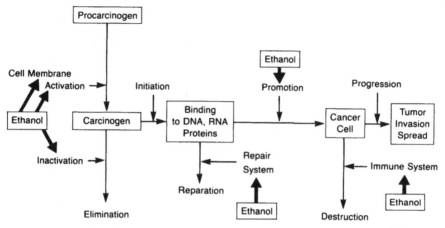

Figure 2. Simplified scheme of two-step carcinogenesis and possible sites of action of ethanol. From Spitz (1985).

treatments enhanced the occurrence of nitrosamine-induced extrahepatic carcinogenesis in the esophagus (Gibel, 1969) or nervous tissue (Griciute et al., 1981).

The basis for the suggestion that ethanol may affect nitrosamine-induced cancer was derived from observations that ethanol competitively inhibits the activity of hepatic low K_m dimethylnitrosamine (DMN) demethylase, the enzyme which bioactivates DMN to an ultimate carcinogenic form.[2] More specifically, Peng et al. (1982) reported a K_i of 0.31 mM in vitro, while Tomera et al. (1984) noted that less than 0.5 mM ethanol inhibited this enzyme in the perfused rat liver. In contrast, chronic consumption of ethanol enhances hepatic cytochrome P-450 and liver microsomal high and low K_m DMN demethylase (Schwartz et al., 1980; Garro et al., 1981). In fact, the enhanced microsomal activity caused by ethanol enhanced the DMN-induced mutagenicity in the Ames test (Garro et al., 1981). According to Seitz and Simanowski (1986), the DMN concentrations used in the Garro et al. (1981) report of less than 0.3 mM may be of pathophysiological importance.[3] Despite suggestive findings for enhanced nitrosamine-induced mutagenicity in the Ames assay, follow-up studies could find no ethanol treatment effect with respect to methylation of hepatic DNA (Schwartz et al., 1982), or when the mutagenicity of DMN was tested in vivo using the host-mediated assay (Glatt et al., 1981). Consequently Seitz and Simanowski (1986) concluded that the generally insignificant effect on nitrosamine-induced hepatic tumor incidence by ethanol is more understandable given these negative findings.

While the alcohol treatment does not apparently enhance the hepatocarcinogenicity of DMN, it does enhance nitrosamine-induced esophageal cancer (Figure 18.3). In addition, administration of DMN and ethanol results in a 1.8- to 4.6-fold rise in the methylation of esophageal DNA (Swann et al., 1984).

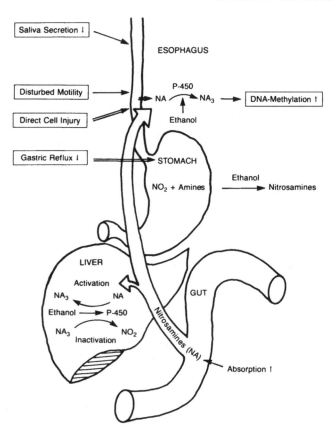

Figure 3. Effect of ethanol on nitrosamine metabolism and esophageal carcinogenesis. Ethanol increases gastric production and intestinal absorption of nitrosamines. Nitrosamine metabolism in the liver is inhibited by ethanol, and therefore extrahepatic tissues such as the esophagus are exposed to higher concentrations of nitrosamines. In the esophageal mucosa, nitrosamine-metabolizing enzymes are induced by alcohol. NA, nitrosamines; NA_3, activated nitrosamines. From Seitz and Simanowski (1986).

Further studies indicate that the alkylation of esophageal DNA is greater than in the liver after a small (versus large) dose of ethanol, probably because of the low K_m of the esophageal metabolizing system compared to the liver and kidney.

Another reason the extrahepatic tissues may be at greater risk than the liver is that the pharmacokinetics of small oral doses of certain nitrosamines such as DMN are dominated by first-pass clearance in the liver (Diaz Gomez et al., 1977; Pegg and Perry, 1981; Driver and Swann, 1987). In fact, oral doses of less than 30 µg/kg are principally removed from the portal blood as it passes through the liver, and practically none of the nitrosamine reaches the general circulation. Consequently extrahepatic organs that can bioactivate the nitrosamine to carcinogenic form(s) — including the kidney and the lung in rats and nasal epithelium in mice — are usually protected.

However, ethanol exposure can have a major impact on this protective mechanism. For example, Driver and Swann (1987) indicate that even small amounts of ethanol (e.g., 10 mL of 5% ethanol/kg) will prevent the first-pass clearance and enhance exposure of the extrahepatic organs to DMN (Swann, 1982; Swann et al., 1984). The prevention of the first-pass clearance results from the competitive inhibition of DMN metabolism. In practical terms, the human equivalent dose based on rat data to inhibit the first-pass clearance effect is the drinking of one pint of beer (Spiegelhalder et al., 1982; Spiegelhalder and Preussmann, 1985). This prevention of the first-pass clearance exposes organs such as the esophagus that typically would not be in contact with small quantities of the nitrosamine coming from the GI tract (Swann, 1982). According to Driver and Swann (1987), the results of cancer bioassays on the joint exposure of ethanol and DMN can be accounted for by the effect of ethanol on first-pass clearance. It is also interesting to note that ethanol enhanced the esophageal carcinogenicity of DMN only if given concurrently but not if given at split time intervals (Teschke et al., 1983). Again, this is consistent with the perspective that the modification in carcinogenicity is caused by a change in the first-pass clearance as a result of the inhibition of hepatic nitrosamine metabolism (Driver and Swann, 1987).

Seitz and Simanowski (1986) noted other possible mechanisms by which ethanol may affect the occurrence of esophageal carcinogenesis. Ethanol is known to diminish esophageal motility and may possibly increase the exposure of the esophageal mucosa to procarcinogens. Ethanol also alters cell membrane fluidity and permeability and may therefore enhance penetration of the procarcinogen into the cell (Shirazi and Platz, 1978; Freund, 1979; Salo, 1983; Kuratsune et al., 1965). Furthermore, alcohol consumption can also decrease the production and secretion of saliva while increasing viscosity, thereby causing higher localized concentrations of procarcinogens and induced cleansing of the mucosal surface (Kissin and Kaley, 1974; Maier et al., 1986). It is also known that alcohol enhances esophageal reflux (Kaufmann and Kaye, 1978), which eventually may result in Barrett's Syndrome, related to an enhanced risk of esophageal adenocarcinoma (Savary et al., 1981). Thus Seitz and Simanowski (1986) concluded that both "systemic and local ethanol-related effects may be responsible for the enhanced esophageal carcinogenesis following chronic ethanol ingestion."

X. Pesticides

In an effort to obtain an initial insight concerning the effects of chronic alcohol ingestion on the toxicity of pesticides, Maita et al. (1988) assessed the capacity of chronic ethanol ingestion to affect the hepatotoxicity of tri-ortho cresyl phosphate (TOCP) and 2,4,6-trichlorophenyl 4 nitrophenyl ether (MO) using F344 male rats. The investigators exposed the rats to a large amount of ethanol in the diet (5 g/100 g) for a three-week period. This was followed by a single p.o. dose of either MO or TOCP. Using serum biochemical parameters as indicators of toxicity, it was found that the ethanol pretreatment appeared to enhance by about two- to fourfold the occurrence of MO-induced liver damage as measured by GOT and GPT, while protecting against similar damage caused by TOCP by about two- to sevenfold. That ethanol enhanced the toxicity of MO and prevented TOCP toxicity was unexpected. Unfortunately, no mechanistic research was available to account for these findings. Clearly, further research is needed in this area.

Y. Styrene

Biological monitoring for styrene as a result of occupational exposure is performed by measuring the excretion of mandelic acid in the urine (Engstrom et al., 1976; Astrand et al., 1974; Wilson et al., 1983; Wolff et al., 1978). The optimum sampling time is now accepted as being at the end of the work shift (Wilson et al., 1983). However, it has been recognized that consumption of moderate levels of alcohol during the course of the workday can significantly inhibit the metabolism of styrene to mandelic acid and result in a false low value (Wilson et al., 1983).

More specifically, in an earlier study Wilson et al. (1979) reported that the excretion of mandelic acid by two technicians building glass-reinforced plastic boats under model ventilation conditions achieved maximum excretion values 4–8 hours after the end of exposure. Since this apparent time lag was independent of exposure, it caused some controversy over when the actual optimum time for biomonitoring should occur. However, since these two technicians had consumed beer during their lunch period at midday, it led to the suggestion that ethanol may have affected the normal rate of styrene metabolism.

Since alcohol consumption was known to affect the metabolism of xylene, toluene, and trichloroethylene (Sato, A., et al., 1980, 1981a; Elovaara et al., 1980; and Muller et al., 1975), it was decided to examine carefully whether modest ethanol consumption may be a factor in affecting styrene metabolism. In their 1983 follow-up study, Wilson et al. assessed the effect of a dose of alcohol (i.e., sufficient to produce a blood alcohol 0.06%) on the kinetics of mandelic acid excretion in four male volunteers aged 39 to 45 exposed to 220 mg/m^3 styrene under controlled exposure chamber conditions. The authors observed that ethanol reduced the excretion of mandelic acid such that the maximum excretion occurred not at the end of the exposure period (as occurred in the non–alcohol exposed controls) but 3 hours after the end of exposure. It should be noted that 1 hour following ethanol administration, the blood mandelic acid levels were only 56% of the controls, and this was accompanied by a 15-fold increased excretion of phenylethane 1,2-diol, the major metabolic precursor of mandelic acid.

This marked change in excretion rates between blood mandelic acid and phenylethane diol strongly supports the premise that the inhibition of styrene metabolism by ethanol occurs at the oxidative conversion of the diol to mandelic acid. From a mechanistic perspective, the authors suggested that the inhibition of the oxidation of this diol is associated with the change in $NAD^+/NADH$ ratio caused by the ethanol metabolism.

Based on their findings, Wilson et al. (1983) concluded that

> spuriously low concentrations of mandelic acid may be obtained as estimates of styrene uptake if there is ethanol ingestion during the working day. The mandelic acid concentration in the "end-of-shift" sample after ethanol was less than 30% of the value obtained in the first non-alcoholic "shift." A discrepancy of this extent is enough to reduce any correlation between uptake and excretion if an unknown proportion of the work-force being investigated is taking alcohol during the study period, and will make it difficult to make judgments concerning individuals in routine occupational health practice.

Z. Tetrahydrocannabinol

In its use as a vehicle, ethanol has been shown by various authors to enhance the absorption of Δ^9 tetrahydrocannabinol (THC) compared to normal saline and Tween 80 (Ho et al., 1971; Sofia et al., 1971, 1974; Paton, 1975). Later studies by Clarke and Jandhyala (1977) revealed that ethanol enhanced the occurrence of Δ^9 THC induced increases in brain MAO levels. Clearly further research is needed in this area.

AA. Toluene

It has been demonstrated in studies with human subjects that ingestion of ethanol diminishes the metabolism of simultaneously inhaled aromatic hydrocarbons including toluene, m-xylene, and styrene (Waldron et al., 1983; Riihimaki et al., 1982a; Wilson et al., 1983). Ethanol impairment of the metabolism of these solvents may be explained by observations that ethanol inhibits enzyme activities involved in their respective degradations (Ikeda and Ohtsuji, 1971; Patel et al., 1978; Pantarotto et al., 1980; Rubin et al., 1970). A. Sato et al. (1980, 1981b) reported dose-related inhibition of a single moderate ethanol dose on the in vitro metabolism of toluene, xylene, styrene, and TCE by microsomal enzymes in the rat liver. However, the metabolic rates of these and other hydrocarbons were increased (Table 18.3) in rats following chronic feeding of ethanol. Interestingly, a one-day withdrawal of ethanol nearly eliminates this enhancing effect of chronic ethanol consumption in the rat.

The relationship of ethanol consumption to metabolism of xenobiotics such as toluene and related agents is complex, having both stimulatory and inhibitory effects depending on exposure conditions. For example, single moderate exposures of ethanol administered to non–occupationally exposed per-

Table 3. Stimulation of Drug-Metabolizing Enzyme Activities After Chronic Ethanol Consumption

Hydrocarbons	Metabolic Rate[a] (nmol/g liver/min)			EtOH(+)/ control[e]
	Control[b]	EtOH(+)[c]	EtOH(−)[d]	
Benzene	13.7 ± 5.4	87.5 ± 13.7	15.5 ± 3.3	6.4
Toluene	18.1 ± 4.9	88.5 ± 2.6	20.5 ± 5.4	4.9
m-Xylene	21.0 ± 5.5	76.6 ± 2.6	22.8 ± 5.1	3.6
Styrene	28.5 ± 4.3	90.3 ± 10.1	30.8 ± 8.3	3.2
Dichloromethane	28.5 ± 1.5	150.0 ± 21.9	36.0 ± 7.2	5.3
Chloroform	19.7 ± 2.6	126.6 ± 10.5	20.2 ± 1.1	6.4
Carbon tetrachloride	1.9 ± 0.2	8.3 ± 1.5	1.7 ± 0.3	4.4
1,1-Dichloroethane	19.1 ± 3.3	120.9 ± 7.5	20.3 ± 0.5	6.3
1,2-Dichloroethane	23.6 ± 1.1	128.6 ± 7.9	22.7 ± 0.3	5.5
1,1,1-Trichloroethane	0.5 ± 0.2	1.8 ± 0.5	0.6 ± 0.3	3.6
1,1,2-Trichloroethane	21.0 ± 1.9	117.6 ± 2.1	21.5 ± 1.3	5.6
1,1,1,2-Tetrachloroethane	8.1 ± 2.0	29.4 ± 5.5	11.0 ± 3.1	3.6
1,1,2,2-Tetrachloroethane	13.3 ± 0.6	70.9 ± 8.3	16.2 ± 2.5	5.3
1,1-Dichloroethylene	31.1 ± 6.6	100.6 ± 10.8	34.1 ± 4.9	3.2
Trichloroethylene	18.9 ± 7.4	105.3 ± 1.5	20.1 ± 5.3	5.6
Tetrachloroethylene	0.5 ± 0.3	2.6 ± 0.4	0.5 ± 0.4	5.2

Source: Sato et al. (1980).
[a]Mean ± SD obtained from five rats.
[b]Rats fed control diet up to the day before death.
[c]Rats fed ethanol-containing diet up to the day before death.
[d]Ethanol-treated rats fed control diet in place of ethanol-containing diet only on the day before death.
[e]Data in the column "Control" have been reported previously.

sons have been found to delay elimination of inhaled toluene at the Swedish TLV of 3.2 mmol/m³ (Wallen et al., 1984). However, a single dose of ethanol (5 g/kg) in rats enhanced the metabolism of toluene and TCE about twofold 16 hours after administration of the ethanol without affecting microsomal protein and cytochrome P-450 (Sato et al., 1980). Thus variable patterns of ethanol consumption are likely to have somewhat different effects on the pharmacokinetics of toluene and related agents. Insufficient research exists to allow adequate prediction of the pharmacokinetic relationships for a wide range of human alcohol consumption patterns.

BB. Tranquilizer Combinations

There is an extensive literature that has assessed the capacity for various tranquilizers to interact with ethanol in animal models and human systems, usually considering behavioral endpoints such as various aspects of central nervous system depression (Stolman, 1969). Such interaction studies have involved ethanol with numerous tranquilizers — chlorpromazine being studied most frequently. A generalization that emerges from binary studies involving at least 18 tranquilizers with ethanol indicates that most tranquilizers display a capacity to enhance the central nervous system effects of ethanol.

Interactions of chlorpromazine with ethanol have been studied in numerous species, including mice, rats, rabbits, dogs, and humans. In general, when chlorpromazine is given prior to ethanol, it enhances a variety of ethanol-induced effects, including sleeping time, immobility, tremors, and unsteadiness. Considerable efforts have attempted to determine temporal relationships between chlorpromazine and ethanol. More specifically, in mice the most effective enhancement of ethanol-induced narcosis by chlorpromazine occurred 1–5 hours following an hypnotically ineffective chlorpromazine dose. The magnitude of the enhancement varied according to the experimental conditions. Nevertheless, Kopf (1957) reported that chlorpromazine enhanced the sleeping time of ethanol-treated mice by nearly 20-fold. The pharmacokinetic nature of the metabolic interaction between chlorpromazine and ethanol is such that the chlorpromazine decreased ethanol metabolism in rabbits, dogs, and humans, as inferred by higher plasma alcohol levels following treatment. The magnitude of the apparent inhibition of ethanol metabolism again varies somewhat depending on the model employed and the experimental conditions but would be in the 15–30% range.

Other tranquilizers that have been shown to enhance the central nervous system depressant effects of ethanol include

- reserpine (Brodie and Shore, 1957; Aston and Cullumbine, 1960)
- promazine (Eerola, 1963)
- promethazine (Phenagran) (Lish et al., 1960; Eerola, 1963)
- hydroxyzine (Atarax) (Forney et al., 1962; Eerola, 1963)
- meprobamate (Aston and Cullumbine, 1960; Forney et al., 1962)
- chlordiazepoxide (Librium) (Zbinden et al., 1961)
- imipramine (tofranil) (Herr et al., 1961)
- methdilazine (Tacaryl) (Lish et al., 1960)
- azacyclonal (Frenquel) (Aston and Cullumbine, 1960)
- phenaglycodol (Ultran) (Forney et al., 1962)
- chlorprothixine (Taractan) (Herr et al., 1961)
- benactyzine (Suavitil) (Holten and Larsen, 1956)
- N-methylpiperidyl-3-methyl phenothiazine (Kopf, 1957)
- benzquinamide (Khan et al., 1964)

These investigations were conducted with mice as the animal model and durations of sleeping time or righting reflex as the endpoint measured. The magnitude of the enhancements varied according to a wide range of experimental conditions. For example, Lish et al. (1960) reported a 138.3% increase in sleeping time when rats were pretreated with 6.18 mg methdilazine/kg i.p. 30 minutes prior to the administration of 23 mL 10% ethanol/kg i.p The magnitude of the enhancement could easily be manipulated by modification of any of one of several critical variables.

A number of different investigators assessed the capacity of ethanol and morphine to interact in rodents. In general, morphine pretreatment markedly enhanced the immobility time of mice in a dose-dependent fashion (Forney et al., 1962). In fact, increasing the dose of morphine by a factor of 8 increased the time of immobility by a similar fashion. Other investigators found that blood levels of ethanol associated with respiratory failure in rats were significantly diminished when given morphine (Haggard et al., 1940). Lethality studies using mice again revealed an enhancement of ethanol toxicity by morphine. In fact, Eerola et al. (1955) reported that the joint effect of ethanol and morphine reflected a clear enhancement, with the magnitude of enhancement being more distinct with smaller than larger doses. For example, in instances where the additive mortality would have been 10–25% with smaller doses, it was nearly threefold greater.

An additive response was reported in lethality studies involving ethanol and acetanilid using rabbits (Higgins and McGuigan, 1983), while no interaction was observed between ethanol and d-proporyphene (Darvon) with respect to immobility in mice (Forney et al., 1962).

CC. 1,1,1-Trichloroethane

In 1981 Woolverton and Balster assessed the acute behavioral and toxicological effects in male mice of the volatile solvent 1,1,1-trichloroethane, alone and in combination with oral ethanol. The rationale for the study was the predicted likelihood that persons using 1,1,1-trichloroethane as a recreational drug may use ethanol as well. When both agents were administered, the ethanol was given 30 minutes prior to 30 minutes exposure to 1,1,1-trichloroethane. With respect to lethality, the combinations of low doses of ethanol with 1,1,1-trichloroethane were usually supra-additive, while higher doses of ethanol were additive or infra-additive with 1,1,1-trichloroethane. As for behavioral responses, effects of low levels of 1,1,1-trichloroethane were supra-additive, while the effects of high concentrations of 1,1,1-trichloroethane were additive or infra-additive.

In conclusion, the authors stated that high doses of exposure to ethanol generally

> reduces both the behaviorally active and lethal concentrations of 1,1,1-trichloroethane in mice, with a supra-additive interaction usually found with low doses of ethanol. It should be noted, however, that additive or infra-additive interactions are also functionally important when dose-response functions are steep as they are with ethanol and 1,1,1-trichloroethane. . . . These data lend support to the suggestion that oral ethanol consumption prior to either voluntary or involuntary exposure to 1,1,1-trichloroethane would often be predicted to present a greater risk than 1,1,1-trichloroethane exposure alone and that combined exposure to these compounds may represent a significant health hazard.

DD. Trichloroethylene

The effects of ethanol administration on trichloroethylene (TCE) metabolism have been studied in humans by Muller et al. (1975), who found that persons exposed simultaneously to 50 ppm TCE for 6 hours and ethanol via ingestion (blood level of 60 mg %) experienced a decrease in the rate of oxidation to trichloroethanol and trichloroacetic acid (TCAA) by 40%. Furthermore, the ethanol-exposed volunteers displayed blood levels of TCE 2.5-fold above control values. Oxidation of TCE to TCAA did not occur if there were detectable levels of ethanol in the blood. Ethanol and acetaldehyde levels were slightly greater than those of the control.

The findings of Muller et al. (1975) were consistent with a subsequent report by White and Carlson (1981), which noted that blood levels of TCE in ethanol-treated rabbits were significantly higher, and TCAA levels significantly lower, than in controls with ethanol exposure. Furthermore, TCE exposure markedly lengthened the blood half-life of ethanol.

Not only did these collective studies reveal a metabolic interaction but White and Carlson (1981) also established that epinephrine-induced cardiac arrhythmias in rabbits exposed to TCE were potentiated by ethanol. While this potentiation was associated with the reduced metabolism of TCE, no clear understanding of the mechanism of action was elucidated.

A comparable enhancement of TCE-induced toxicity was reported by Utesch et al. (1981), who assessed the effects of this combination exposure on CNS depression in Sprague-Dawley rats. Likewise, Ferguson and Vernon (1970) reported that ethanol enhanced the inhibitory effects of 300 and 1000 ppm TCE on visual-motor performance in humans.

A key feature of the nature of the interaction of ethanol and TCE is temporal in nature. Ethanol dehydrogenase converts ethanol to acetaldehyde, but it also converts a TCE metabolite, chloral hydrate, to trichloroethanol. Thus, when the two agents are administered simultaneously, it is evident that ethanol will have a direct impact on TCE metabolism. However, predosing with ethanol may have an opposite effect on TCE metabolism since a single dose of ethanol, given 17–24 hours prior to the second drug, will enhance liver microsomal metabolism and the plasma elimination of the second agent (Mallov and Baesl, 1972; Powis, 1975; Sato et al., 1980). Since TCE is bioactivated to a toxic intermediate by the MFO system, the enhancement of its metabolism by ethanol has also resulted in the potentiation of TCE-induced liver toxicity in male Sprague-Dawley rats (Cornish and Adefuin, 1966), although not in the Utesch et al. (1981) rat study.

EE. Vinyl Chloride

It is now well accepted that chronic inhalation of high concentrations of vinyl chloride for many years is causally associated with an increased risk of angiosarcoma of the liver in humans (Creech and Johnson, 1974; Delorme and Mark, 1975) and animal models (Viola et al., 1971; Maltoni and Lefemine, 1974). Vinyl chloride has been shown to be metabolized via the alcohol dehydrogenase pathway to 2-chloroethane, chloroacetaldehyde, and monochloroacetic acid. It is believed that only minute quantities, if any, of the monochloroacetic acid are found at low doses since chloroacetaldehyde reacts quickly with both glutathione and cysteine (Johnson, 1967). Hefner et al. (1975) have proposed that this conjugation explains the lack of dose-related reactions in nonprotein sulfhydryl content of the rat liver when the rats are exposed to 50–15,000 ppm of vinyl chloride.

The alcohol dehydrogenase pathway for vinyl chloride, however, becomes saturable at relatively low levels (220 ppm), and alternative pathways are believed to become involved in the metabolism of vinyl chloride at high exposure levels. Some of the alternative pathways have been speculated to involve the oxidative transformation of vinyl chloride to chloroethylene oxide, which itself spontaneously rearranges to chloroacetaldehyde. This interpretation was supported by observations that SKF-525A causes some inhibition of vinyl chloride metabolism in rats exposed at 1038 ppm, but not 65 ppm. Hefner et al. (1975) also argued that this hypothesis may account for why monochloroacetic acid may be excreted by rats exposed at 5000 ppm, but not at 50 ppm.

The quest to clarify the toxicodynamics of vinyl chloride may be of considerable significance since "the saturation of a primary metabolic pathway for vinyl chloride monomer degradation and redirection through other pathways provide some hope that a threshold concentration for the untoward effects of vinyl chloride monomer may exist. Metabolites of vinyl chloride monomer formed only via the alternate pathways may constitute the ultimate toxin and carcinogen" (Hefner et al., 1975). For example, the postulated formation of chloroethylene oxide, a very active bifunctional alkylating agent, is particularly relevant to such a discussion.

Of particular interest is that the administration of pyrazole, an inhibitor of alcohol dehydrogenase, xanthine oxidase, and other enzymes (Carter and Isselbacher, 1972), inhibits the metabolism of vinyl chloride, suggesting that metabolism of vinyl chloride is via alcohol dehydrogenase. Supportive evidence for the metabolism of vinyl chloride by alcohol dehydrogenase was its inhibition by the administration of ethanol (Hefner et al., 1975; Hultmark et al., 1979).

As noted above, vinyl chloride is metabolized via the alcohol dehydrogenase pathway, saturable at high concentrations. Furthermore, alternative pathways have been proposed (Hefner et al., 1975) as forming possible proximate carcinogenic forms. Consequently it may be speculated that consumption of sufficient ethanol may divert vinyl chloride metabolism to alternative pathways and modify cancer risk. In 1981 Radike et al. assessed the effect of ethanol on vinyl chloride–induced carcinogenesis in male Sprague-Dawley rats. In this study the rats were exposed via inhalation with 600 ppm vinyl chloride 4 hours/day, 5 days/week, for 1 year. Ethanol was administered with the drinking water at 5%. Terminal sacrifice occurred 2.5 years after the start of vinyl chloride exposure.

Consistent with the theory, the results indicated that ethanol potentiated the occurrence of vinyl chloride–induced liver cancer, and there was no excess of tumors in the animals receiving ethanol alone. More specifically, the combined ethanol vinyl chloride treatment displayed a 50% incidence of angiosarcomas, compared to 23% in the vinyl chloride–treated group. The combined treatments also enhanced the occurrence of hepatocellular carcinoma and lymphosarcoma.

FF. Vitamin A

Both vitamin A and ethanol ingestion in excess for prolonged periods cause fatty infiltration of the liver. Consequently it was of interest to assess the impact of these two agents when they were applied jointly at high levels of exposure for a prolonged period. Researchers exposed weanling male Sprague-Dawley rats to a diet that contained either about 400 IU of retinyl acetate/day or 5 times that amount. Note that this is a quantity that itself is not hepatotoxic. These animals also received 36% of calories as ethanol, replacing other carbohydrate components of the diet. The exposure groups consisted of vitamin A–supplemented, ethanol-fed, and vitamin A–supplemented plus ethanol-fed; control groups were pair-fed the basic diet. The animals were sacrificed after two and nine months (Leo et al., 1982; Leo and Lieber, 1983).

After two months, significant fatty infiltration of the liver had occurred in the ethanol plus vitamin A group, with only moderate steatosis in the livers of rats consuming the alcohol without vitamin A supplement. After nine months the morphological changes in the liver became considerably more extensive in the vitamin A supplement–alcohol group.

The authors concluded that ingestion of a nontoxic amount of vitamin A can enhance the hepatotoxicity of ethanol in rats. The morphological changes involved inflammation, increased numbers and size of lipocytes, hepatocellular necrosis and fibrosis, alterations in mitochondrial function, and changes in serum enzymes levels. Whether these findings can be directly extrapolated to humans is uncertain. However, it is known that supplements of vitamin A of similar magnitude to those used in this study are seen in the treatment of biliary cirrhosis (Herlong et al., 1981), of abnormal dark adaptation in alcoholic patients (Russell et al., 1978), and of a variety of skin disorders (Fleischmann et al., 1977). Consequently their findings are of considerable potential clinical importance to evaluate and extend.

GG. Xylene

The metabolism of xylene involves the microsomal oxidation of one of the methyl groups (major metabolic pathway) and of the aromatic ring (minor metabolic pathway). The side chain oxidation yields alcoholic and aldehyde intermediates, which are catalyzed by alcohol and aldehyde dehydrogenases to the corresponding organic acid, which becomes conjugated and ultimately excreted (Riihimaki et al., 1982a). Since the metabolism of xylene therefore is contingent on the alcohol and aldehyde dehydrogenase catalyzed reactions,

Riihimaki et al. (1982a) hypothesized that ethanol and xylene may interact metabolically. In their study, Riihimaki et al. (1982a) found that ingestion of a moderate dose of ethanol (0.8 g/kg) by human volunteers prior to a 4-hour inhalation exposure to m-xylene (6.0 or 11.5 mmol/m^3) caused significant alterations in xylene kinetics. More specifically, following the ethanol administration the blood xylene levels rose about 1.5-to 2.0-fold, while urinary methylhippuric acid excretion, the principal excretory metabolite, declined by about 50%. Such findings suggest that ethanol decreased the metabolic clearance of xylene by about 50% during xylene inhalation. The duration of the reduced excretion persisted for several hours.

Riihimaki et al. (1982b) indicated that these findings closely resemble the well-known TCE–ethanol interaction. In that case TCE undergoes microsomal oxidation, forming chloral hydrate, which is further converted to trichloro-ethanol by alcohol dehydrogenase (Leibman and Ortiz, 1977). Since ethanol exposure is known to inhibit microsomal drug metabolism (Rubin et al., 1970; Rubin and Lieber, 1968) as well as to competitively inhibit alcohol dehydro-genase, as previously seen with respect to methanol (Makar et al., 1968) and ethylene glycol (Wacker et al., 1965; Peterson et al., 1968), it is likely that the interaction of ethanol and xylene may be complex, at least affecting several enzyme systems.

While the prior discussion considered the effects of ethanol on xylene metabolism, other experiments by Riihimaki et al. (1982a,b) also observed that m-xylene inhalation appeared to affect ethanol metabolism, as evidenced by increased concentrations of blood acetaldehyde. While the toxicological impli-cations of the xylene–ethanol interaction remain to be explored, Riihimaki et al. (1982a,b) suggest that the interaction can be manifested as dizziness, nau-sea, or dermal flush.

HH. Other: Aldehyde Dehydrogenase Inhibitors

A variety of agents have been found to inhibit the enzyme aldehyde dehydrogenase (ADH). Recognition of this property has led to the develop-ment of a therapeutic approach for the treatment of alcoholism using, for example, disulfiram (Antabuse) or calcium carbamide (Temposil) (Hillbom et al., 1983). However, it is now recognized that a number of agents used in industry or pharmaceutically as medicinal agents also inhibit aldehyde dehy-drogenase, including tetramethylthiuram disulfide (TMTD), a rubber accelera-tor and fungicide; tetramethylthiuram monosulfide (TMTM), a rubber accel-erator; zinc dimethyldithiocarbamate (Ziram), a fungicide; zinc ethylenebis (dithiocarbamate) (Zineb); Ca cyanamide (Cyanamide), a fertilizer and drug (Garcia de Torres et al., 1983); dimethylformamide (Lyle et al., 1979); chlorpropamide (*Br. Med. J.*, 1964); pyrazole; 4-methylpyrazole; and 3-amino-1,2,4-triazole (3AT) (Phillips et al., 1977).

It has been speculated that since a variety of important environmental contaminants are metabolized by the enzyme aldehyde dehydrogenase, including alcohol, their toxicity may be modified considerably by alcohol ingestion or other inhibitors of ADH (see Chapter 20). Limited studies have shown, for example, that pyrazole, 4-methyl pyrazole, 3-AT, and disulfiram markedly affect the metabolism of the carcinogen dimethylnitrosamine (Phillips et al., 1977). Given the impressive toxicological database associated with disulfiram, a model aldehyde dehydrogenase inhibitory compound, it suggests that other aldehyde dehydrogenase inhibitors may likewise display a significant potential to interact with agents of toxicological concern. Clearly, more research is needed in this area.

IV. SUMMARY

That ethanol may affect the toxicity of inorganic agents has been demonstrated with cadmium, cobalt, lead, mercury, and H_2S. In the cases of cadmium and mercury, the ethanol treatment was found to be protective, while it enhanced toxicity of cobalt, lead, and H_2S. The mechanisms of action for various interactions are remarkably varied, ranging from increasing the metallothionein concentration in the liver and thereby reducing cadmium toxicity, to inhibiting the oxidation of inhaled Hg^0 to Hg^{++} and thereby facilitating its excretion via the lungs. The inorganic contaminant ethanol interaction database is rather modest in comparison to studies with organic agents and is based principally on studies with predictive animal models. Nevertheless, the findings in this section, especially those relating to lead, have important public health implications and are worthy of epidemiologic consideration.

This chapter has unequivocally established that exposure to ethanol can play a significant role in the expression of toxicity and carcinogenicity of numerous organic contaminants (Table 18.4). For example, animal studies have revealed that ethanol exposure enhances the occurrence of both vinyl chloride– and nitrosamine-induced carcinogenesis, and the underlying mechanisms are reasonably understood. The amount of ethanol required to affect xenobiotic mechanisms is amazingly low, with about a can of beer being sufficient to alter first-pass kinetics in nitrosamines, resulting in higher nitrosamine exposure in extrahepatic tissue, and to a marked slow down in toluene metabolism and excretion.

The range of agents that ethanol affects, as well as its impressive mechanistic diversity in affecting toxic interactions, is perhaps its most extraordinary feature. Given the common nature of ethanol ingestion in human activities and its impressive capacity to modify xenobiotic metabolism, including bioactivation/detoxification, greater attention needs to be directed toward developing a clearer picture of its role in the development of environmentally related disease.

Table 4. Summary of the Effects of Ethanol on Drug/Pollutant Toxicity

Agent	Nature of Studies/ Types of Interaction	Mechanism	Comment
Inorganics			
Cadmium	Ethanol diminished Cd acute toxicity in rats	Ethanol-induced synthesis of hepatic metallothionein; it results in the sequestering of Cd and decreases its accumulation in critical organelles and proteins	Very limited database; exposures were extremely high; uncertain human relevance
Cobalt	Ethanol enhances cobalt toxicity in mice	Ethanol enhances tissue NADH/NAD ratio while cobalt irreversibly complexes with the SH group of dihydrolipoic acid and -keto acid oxidase	Animal studies provide a solid foundation to explain how cobalt-fortified beer caused a special form of heart disease in chronic beer drinkers; this interaction represents a possible supraadditive response in humans in light of animal data
Hydrogen sulfide	Ethanol enhanced the acute toxicity of rats to H_2S	Mechanism unknown	Important industrial hygiene implications; however, the database is extremely limited
Lead	Ethanol enhances orally administered lead body burden, especially in renal tissue in rats	Mechanism unknown	Limited animal data help to explain numerous instances where humans apparently had enhanced lead toxicity associated with consumption of ethanol
Mercury	Animal model and human investigations indicate the ethanol enhances mercury excretion	Ethanol prevents oxidation of Hg to a more lipophilic chemical state	There is a solid database that ethanol reduces Hg body burden; industrial hygiene implications relate to monitoring strategies
Nitrogen dioxide	Limited studies suggest a possible interaction on atrial response to calcium	Mechanism unknown	Database too limited for any firm conclusions
Organics			
Aflatoxin	Ethanol enhances aflatoxin-induced liver toxicity	Uncertain but believed to involve bioactivation	Relevance at human exposure levels is uncertain

Table 4, continued

Agent	Nature of Studies/ Types of Interaction	Mechanism	Comment
Aspirin	Ethanol enhances gastrointestinal toxicity	Both agents can cause gastritis and damage to G.I. tract	This is considered an interaction of considerable clinical relevance as well as one that is frequently overlooked
Barbiturates	Ethanol enhances toxicity of a wide range of barbiturates in both animal models and humans	Multiple mechanisms have been suggested, including both pharmacokinetic and dynamic	This is perhaps the interaction with the largest database and broadest biomedical recognition
Benzene	Ethanol enhanced benzene-induced hematotoxicity	Ethanol is believed to enhance the bioactivation of benzene to highly toxic metabolites	Strong animal model database; information should be of value in epidemiological studies
Acetaminophen	Chronic ethanol exposure enhances liver damage caused by acetaminophen while acute exposure is protective	Chronic exposure to ethanol enhances bioactivation of acetaminophen while acute exposure to ethanol reduces acetaminophen bioactivation	Relevance to humans needs to be better elucidated
Benzo(a)pyrene	Ethanol enhanced the capacity of BP to cause mutagens	Ethanol enhances enzymatically induced bioactivation of BP	Relevance to humans remains to be developed
Benzodiazepines (BDZ)	Considerable animal model data indicates that ethanol increases the capacity of BDZs to affect a variety of CNS endpoints	Ethanol appears to affect the binding of BZD to receptors; ethanol also affects numerous pharmacokinetic parameters including absorption and metabolism of BZD	Clinical implications relate most directly to impacts on neuromotor coordination and impairment of normal function
Caffeine	Caffeine appears to antagonize the acutely toxic effect of ethanol	Mechanism uncertain	Relevance is questionable since extraordinarily high doses of caffeine appear necessary

Table 4, continued

Agent	Nature of Studies/ Types of Interaction	Mechanism	Comment
Carbon disulfide	Ethanol (and many other alcohols) enhance the toxicity of CS_2 in animal models	Ethanol appears to increase CS_2 toxicity by enhancing its bioactivation and retention in critical tissues such as brain	Solid animal model database; however, application to quantitative risk assessment in industrial hygiene is not developed
Carbon monoxide	Limited animal data suggest that high ethanol exposures enhance CO-induced lethality	Mechanism unknown	Important implications for humans involved in fires
Carbon tetrachloride	Ethanol and numerous alcohols enhance the toxicity of CCl_4 both in animal models and in industrial experience	Ethanol enhances the toxicity of CCl_4 by increasing its bioactivation	Industrial hygiene implications are important
Chloral hydrate	Animal model and human data indicate that ethanol enhances chloral hydrate toxicity; animal studies suggest an additive relationship	Mechanistic studies suggest that ethanol alters normal chloral hydrate metabolism leading to greater accumulation of trichloroethanol and enhanced toxicity	Industrial hygiene and medical implications are prominent
Chloroform	Extensive animal model data have confirmed earlier industrial hygiene data that ethanol enhances chloroform toxicity	Bioactivation of chloroform via the MFO is the principal mechanism	Extremely strong database; industrial and environmental implications remain to be more fully addressed
Cocaine	Ethanol affects both acute and latent cocaine toxicity	Mechanism believed to involve activation via MFO	Given widespread use of both drugs, this interaction requires greater assessment
Dimethylhydrazine	Ethanol was not found to enhance colonic tumors caused by dimethylhydrazine	Mechanism unknown	Data suggested a possible protective effect
DMSO	Ethanol alters DMSO metabolism and toxicity	Mechanism unknown	Limited database but worthy of investigation

Table 4, continued

Agent	Nature of Studies/ Types of Interaction	Mechanism	Comment
Dinitrotoluene	Limited occupational data suggest that ethanol enhances toxicity	Mechanism unknown	Very limited and old data base
Ethylene glycol	Both animal and human data indicate that ethanol protects against ethylene glycol-induced toxicity	Ethanol inhibits the metabolism of ethylene glycol to its toxic metabolite	Strong database
Nitrosamines	Ethanol increase extrahepatic cancers caused by nitrosamines	Ethanol alters the tissue distribution of nitrosamines by inhibiting its first-pass effect	Interactions with significant public health implications
Styrene	Modest ethanol ingestion markedly slows styrene metabolism in humans	See Comment	Styrene is metabolized to medelic acid and this is used to monitor styrene exposure; unless ethanol ingestion is considered, faulty interpretation of exposure data is likely
Tetrahydrocannabinol (THC)	Ethanol enhances the absorption of THC	Mechanism unknown	Limited datadase but very high social relevance
Toluene, Xylene	Research with humans indicates that acute ethanol exposure diminishes the metabolism to toluene, xylene; chronic ethanol exposure enhances metabolism	Mechanism appears mediated via inhibition or stimulation of MFOs	Application to industrial hygiene practice requires development
1,1,1-Trichloroethane (1,1,1-TCE)	Ethanol and 1,1,1-TCE displayed additive effects as measured by acute toxicity	Mechanism unknown	Industrial hygiene implications need to be assessed
Trichloroethylene	Ethanol consumption decreases the metabolism of TCE	Mechanism appears mediated via inhibition of MFO	Given the widespread occurrence of TCE contamination, these findings warrant considerable further research
Vinyl chloride	Ethanol enhanced the occurrence of vinyl chloride-induced carcinogenicity	Ethanol competitively inhibits vinyl chloride metabolism; forces vinyl chloride to be metabolized via an alternative pathway	Implications for human cancer risk assessment potentially important

NOTES

1. To assess the ED_{50} or LD_{50} of binary drug mixtures, two agents were administered in various ways in which the dose of one agent of the mixture was fixed (i.e., fixed-dose method) or the ratio of the two agents in the mixture was kept constant (i.e., fixed-ratio method).

2. Dimethylnitrosamine requires bioactivation to both a mutagenic and a carcinogenic form via microsomal-dependent oxidative demethylation. This activation, according to Garro et al. (1981), responds in an inducer-specific manner. More specifically, inducers such as phenobarbital (PB), 3-MCA, and Aroclor 1254 increase the activity of DMN demethylase but only when high DMN concentrations (10–100 mM) are used. In contrast, when concentrations of 4 mM or less have been used, these inducers either have no effect or actually decrease DMN demethylase activity.

   These concentration-dependent effects on DMN demethylase by PB and 3-MCA have been explained by the presence of several isoenzymes of the demethylase (Arcos et al., 1977). In fact, kinetic evaluations have led to the suggestion that there are at least 3 isoenzymes in rats, with K_ms of approximately 40, 1.5, and 0.2 mM (Lake et al., 1974; Venkatesan et al., 1970). Because it is not likely that intracellular DMN concentrations will normally even closely achieve 40 mM, Garro et al. (1981) asserted that it is the lower K_m enzymes, which are noninducible by PB or 3-MCA, that are especially important with respect to the bioactivation of DMN to a mutagen and carcinogen. In contrast to PB and 3-MCA, chronic ethanol ingestion in the rat induces a form of cytochrome P-450 that is distinct from that induced by PB or 3-MCA. Furthermore, chronic ethanol ingestion causes an increase in microsomal DMN demethylase activity, which is detectable over a concentration range of 1 to 100 mM DMN, and an increased activation of DMN to a mutagen at levels less than 1 mM. The findings of Garro et al. (1981) are consistent with previous results, which showed that DMN demethylase activity is also increased in chow-fed rats and mice after the acute administration of ethanol, isopropyl alcohol, or acetone (Maling et al., 1975; Sipes et al., 1973a, 1978).

3. It should be noted that chronic ethanol ingestion has also been found to enhance the bioactivation of procarcinogens to mutagens for benzo(a)pyrene (Seitz et al., 1978, 1981a), 2-aminofluorene (Seitz et al., 1981a), tryptophane pyrolysate (Seitz et al., 1981a), and aflatoxin B_1 (Obidoa and Okolo, 1979). It has been speculated that this ethanol-enhanced activation of BAP may help explain the increased incidence of cancer when BAP is administered systemically (Capel et al., 1978).

REFERENCES

Adams, W.L. (1943). The comparative toxicity of chloral alcoholate and chloral hydrate. *J. Pharmacol. Exp. Therap.* 78:340–345.

Allegri, A. (1935). Cocaina, alcool, dinitrofenolo e blu di metilene nella intossicazione sperimentale da barbiturici. *Boll. Soc. Ital. Biol. Sper.* 10:48–51.

Alstott, R.L., Tarrant, M.E., and Forney, R.B. (1973). The acute toxicities of l-methylxanthine, ethanol, and l-methylxanthine/ethanol combinations in the mouse. *Toxicol. Appl. Pharmacol.* 24:393–404.

Altomare, E., Leo, M.A., Sato, C., Vendemiale, G., and Lieber, C.S. (1984). Interaction of ethanol with acetaminophen metabolism in the baboon. *Biochem. Pharmacol.* 33:2207–2212.

Amsel, L.P., and Levy, G. (1969). Drug biotransformation interactions in man. II. A pharmacokinetic study of the simultaneous conjugation of benzoic and salicylic acids with glycine. *J. Pharm. Sci.* 58:321.

———. (1970). Effect of ethanol on the conjugation of benzoate and salicylate with glycine in man. *Proc. Sci. Exp. Biol. Med.* 135:813–816.

Ann. Intern. Med. (1986). Acetaminophen, alcohol, and cytochrome P-450. 104(3):427–428.

Antal, M., Bedo, M., Constantinovits, G., Nagy, K., and Szepvolgyi, J. (1979). Studies on the interaction of methomyl and ethanol in rats. *Fd. Cosmet. Toxicol.* 12:333–338.

Arcos, J., Davies, D., Brown, C., and Argus, M.F. (1977). Repressible and inducible forms of dimethylnitrosamine-demethylase. *Z. Krebsforsch.* 89:181–199.

Aston, R., and Cullumbine, H. (1959). Studies on the nature of the joint action of ethanol and barbiturates. *Toxicol.* 1:65–72.

———. (1960). The effects of combination of ataraxics with hypnotics, LSD and iproniazid in the mouse. *Arch. Intern. Pharm.* 126:219–227.

Astrand, I. (1975). Uptake of solvents in the blood and tissues of man. A review. *Scand. J. Work Environ. Health* 1:199–218.

Astrand, I., Kilbom, A., Ovrum, P., Wahlberg, I., and Vesterberg, O. (1974). Exposure to styrene. I. Concentration in alveolar air and blood at rest and during exercise and metabolism. *Work Environ. Health* 11:69–85.

Baarson, K.A., Synder, C.A. Green, J.D., Sellakumar, A., Goldstein, B.D., and Albert, R.E. (1982). The hematotoxic effects of inhaled benzene in peripheral blood, bone marrow, and spleen cells are increased by ingested ethanol. *Toxicol. Appl. Pharmacol.* 64:393–404.

Baker, J.D., Jr., DeCarle, D.J. and Anuras, S. (1977). Chronic excessive acetaminophen use and liver damage. *Ann Intern Med.* 87:299.

Bartlett, G.R. (1952). Mechanism of action of monofluoroethanol. *J. Pharmacol. Exp. Therap.* 106:464–467.

Bates, M.W., Krebs, H.A., and Williamson, D.H. (1968). Turnover rates of ketone bodies in normal, starved and alloxan-diabetic rats. *Biochem. J.* 110:655–661.

Beauchamp, R.O., Bus, J.S., Popp, J.A., Boreiko, C.J., and Goldberg, L. (1983). A critical review of the literature on carbon disulfide toxicity. *CRC Crit. Rev. Toxicol.* 2:169–278.

Beck, J.F., Cormier, F., and Donini, J.C. (1979). The combined toxicity of ethanol and hydrogen sulfide. *Tox. Lett.* 3:311–313.

Bell, R.G. (1956). A new drug for alcoholism treatment. III. Clinical trial of citrated calcium carbamide. *Can. Med. Assoc. J.* 74:797–798.

Bernheim, F., and Handler, P. (1941). Oxidation of some substituted alcohols by rat liver. *Proc. Soc. Exp. Biol. Med.* 46:470–471.

Blair, A.H., and Vallee, B.L. (1966). Some catalytic properties of human liver alcohol dehydrogenase. *Biochem.* 5:2026–34.

Bock, O.A.A. (1976). Alcohol, aspirin, depression, smoking and stress and the patient with a gastric ulcer. *S. Afr. Med. J.* 50:293–297.

Boost, G. (1965). Discussion remark. In *Dimethylsulfoxyd-DMSO, Schering-Symposium 1965*, G. Laudahn and H.J. Schlosshauer, Eds. E. Blaschker Verlag, Berlin, West Germany.

Branchflower, R.V, and Pohl, L.R. (1981). Investigation of the mechanism of the potentiation of chloroform-induced hepatotoxicity and nephrotoxicity by methyl n-butyl ketone. *Toxicol. Appl. Pharmacol.* 61:407–413.

Brigden, W. (1957). Uncommon myocardial diseases. *Lancet* 2:1173.

Br. Med. J. (1964). Alcohol sensitivity to sulphonyureas. 2:586–587.

Brodie, B.B., and Shore, P.A. (1957). A concept for a role of serotonin and nor-epinephrine as chemical mediators in the brain. *Ann. NY Acad. Sci.* 66:631–642.

Brown, B.R., Jr. (1972). Hepatic microsomal lipoperoxidation and inhalation anesthetics: a biochemical and morphological study in the rat. *Anesthesiology* 36:458–465.

Brown, B.R., Jr., Sipes, I.G., and Sagalyn, A.M. (1974). Mechanisms of acute hepatic toxicity: chloroform, halothane and glutathione. *Anesthesiology* 41:554–561.

Brown, E.M., and Hewitt, W.R. (1984). Dose-response relationships in ketone-induced potentiation of chloroform hepato- and nephrotoxicity. *Toxicol. Appl. Pharmacol.* 76:437–453.

Brown, R.K., and Mitchell, N. (1956). The influence of some of the salicyl compounds (and alcoholic beverages) on the natural history of peptic ulcer. *Gastroenterology* 31:198–203.

Burbridge, T.N., Tipton, D., Sutherland, V.C., and Simone, A. (1958). Effect of chlorpromazine on blood alcohol level. *Fed. Proc.* 17:355:

Burrows, E.H.S. (1953). Alcohol–barbiturate synergism. *S. Afr. Med. J.* 27:1057–1059.

Cabana, B.E., and Gessner, P.K. (1970). The kinetics of chloral hydrate metabolism in mice and the effect thereon of ethanol. *J. Pharmacol. Exp. Therap.* 174:260–275.

Capel, I.D., Turner, M., Pinnock, M.H., and Williams, D.C. (1978). The effect of chronic alcohol intake upon the hepatic microsomal carcinogen activation system. *Oncology* 35:168–170.

Cardani, A., and Farina, G. (1972). Influence of alcoholic beverages consumption on lead-induced changes of haeme biosynthesis. *Med. Lav.* 63:22–28.

Carpenter, C.P., Geary, D.L., Myers, R.C., Nachreiner, D.J., Sullivan, L.J., and King, J.M. (1977). Petroleum hydrocarbon toxicity studies. XV. Animal responses to vapors of "high nephthenic solvent." *Toxicol. Appl. Pharmacol.* 41:251–260.

Carpenter, C.P., Kincaid, E.R., Geary, G.L., Sullivan, L.J., and King, J.M. (1975). Petroleum hydrocarbon toxicity studies. VI. Animal and human responses to vapors of "60-solvent." *Toxicol. Appl. Pharamcol.* 34:374–394.

Carriere, G., Huriez, C., and Williquet, P. (1934). Etude experimentale des injections intraveineuses d'alcool au cours d'intoxications par le Gardenal. *Compt. Rend. Soc. Biol.* 116:118-190.

Carter, E.A., and Isselbacher, K.J. (1972). Hepatic microsomal ethanol oxidation. Mechanism and physiologic significance. *Lab. Invest.* 27:283-286.

Chambers, C.D., and Hunt, L.G. (1977). In *Drug Abuse in Pregnancy and Neonatal Effects,* J.L. Rementeria, Ed., C.V. Mosby, St. Louis, pp. 73-81.

Chan, A.W.K., Greizerstein, H.B., and Strauss, W. (1982). Alcohol–chlordiazepoxide interaction. *Pharmacol. Biochem. Behav.* 17:141-145.

Chan, A.W.K., and Heubusch, P.H. (1982). Relationship of brain cyclic nucleotide levels and the interaction of ethanol with chlordiazepoxide. *Biochem. Pharmacol.* 31:85-89.

Chan, A.W.K., Schanley, D.L., and Strauss, W. (1979). *Res. Commun. Psychol. Psychiat. Behav.* 4:277.

Charbonneau, M., Brodeur, J., DuSouich, P., and Plaa, G.L. (1986a). Correlation between acetone-potentiated CCl_4-induced liver injury and blood concentrations after inhalation or oral administration. *Toxicol. Appl. Pharmacol.* 84:286-294.

Charbonneau, M., Oleskevich, S., Brodeur, J., and Plaa, G.L. (1986b). Acetone potentiation of rat liver injury induced by trichloroethylene–carbon tetrachloride mixtures. *Fund. Appl. Toxicol.* 6:654-661.

Chergelis, C.P., and Neal, R.A. (1980). Studies of carbonyl sulfide toxicity: metabolism by carbonic anhydrase. *Toxicol. Appl. Pharmacol.* 55:198-202.

Cherrick, G.R., and Leevy, C.M. (1965). The effect of ethanol metabolism on levels of oxidized and reduced nicotinamide adenine dinucleotide in liver, kidney, and heart. *Biochem. Biophys. Acta* 107:29.

Childs, A.W., and Lieberman, A.H. (1964). Effect of ethanol on formation of hippuric acid by the liver. *Proc. Soc. Exp. Biol. Med.* 116:881.

Chin, L., Sievers, M.L., Herrier, R.N., and Picchioni, A.L. (1981). Potentiation of pyridoxine by depressants and anticonvulsants in the treatment of acute isoniazid intoxication in dogs. *Toxicol. Appl. Pharmacol.* 58:504-509.

Clark, H., and Powis, G. (1974). Effect of acetone administered in vivo upon hepatic microsomal drug metabolizing activity in the rat. *Biochem. Pharmacol.* 23:1015-1019.

Clarke, D.E., and Jandhyala, B. (1977). Acute and chronic effects of tetrahydrocannabinols on monoamine oxidase activity: possible vehicle/tetrahydrocannabinol interactions. *Res. Comm. Chem. Path. Pharmacol.* 17(3):471-480.

Cornish, H.H., and Adefuin, J. (1966). Ethanol potentiation of halogenated aliphatic solvent toxicity. *Am. Ind. Hyg. Assoc. J.* 27:57-61.

———. (1967). Potentiation of carbon tetrachloride toxicity by aliphatic alcohols. *Arch. Environ. Health* 14:447-449.

Cramer, K. (1966). Predisposing factors for lead poisoning. *Acta. Med. Scand.* 179(suppl.):56-59.

Creech, J.L., Jr., and Johnson, M.N. (1974). Angiosarcoma of liver in the manufacture of polyvinyl chloride. *J. Occup. Med.* 16:150-151.

Curry, S.H., and Smith, C. (1979). Diazepam–ethanol interaction in humans: addiction or potentiation? *Commun. Psychopharmacol.* 3:101-113.

Cushny, A.R. (1924). *Pharmacology and Therapeutics.* Lea and Febiger, Philadelphia, p. 296.

Dalvi, R.R., Hunter, A.L., and Neal, R.A. (1975). Toxicological implications of the mixed-function oxidase catalyzed metabolism of carbon disulfide. *Chem. Biol. Interact.* 10:347–361.

Dalvi, R.R., Poore, R.E., and Neal, R.A. (1974). Studies of the metabolism of carbon disulfide by rat liver microsomes. *Life Sci.* 14:1785–1796.

Danechmand, L., Casier, H., Hebbehunck M., et al. (1967). Combined effects of ethanol and psychotropic drugs on muscular tone in mice. *Q. J. Stud. Alcohol* 28:424–429.

Davidson, D.G.D., and Eastham, W.N. (1966). Acute liver necrosis following overdose of paracetamol. *Br. Med. J.* 5512:497–499.

Davis, W.C., and Ticku, M.K. (1981). Ethanol enhances [3H]diazepam binding at the benzodiazepine gamma-aminobutyric acid receptor onophore complex. *Mol. Pharmacol.* 20:287–94.

De Lamirande, E., and Plaa, G.L. (1981). 1,3-butanediol pre-treatment on the cholestasis induced in rats by manganese-bilirubin combination, taurolithocholic acid, or α naphthylisothiocyanate. *Toxicol. Appl. Pharmacol.* 59:467–475.

Delorme, F., and Mark, L. (1975). Angiosarcomas of the liver in workers having had prolonged contact with vinyl chloride: morphological description of the lesions. *Union. Med. Can.* 104:1836.

Derr, R.F., Aaker, H., Alexander, C.S., and Nagasawa, H.T. (1969). Synergism between cobalt and ethanol on rat growth rate. *J. Nutrit.* 100:521–524.

Desmond, P.V., Patwardhan, R.V., Schenker, S., and Hoyumpa, A.M. (1980). Short term ethanol administration impairs the elimination of chlordiazepoxide (Librium) in man. *Eur. J. Clin. Pharmacol.* 18:275–278.

Diaz Gomez, M.I., Swann, P.F., and Magee, P.N. (1977). The absorption and metabolism in rats of small oral doses of dimethylnitrosamine. *Biochem. J.* 164:497–500.

Dille, J.M., and Ahlquist, R.P. (1937). The synergism of ethyl alcohol and sodium pentobarbital. *J. Pharmacol. Exp. Therap.* 61:385.

Dille, J.M., Linegar, C.R., and Koppanyi, T. (1935). Studies on barbiturates. XII. Factors governing the distribution of barbiturates. *J. Pharmacol. Exp. Therap.* 55:46–61.

Divoll, M., and Greenblatt, D.J. (1981). Alcohol does not enhance diazepam absorption. *Pharmacology* 22:263–268.

Doenicke, A. (1962). Beeintrachtigung der Verkehrssicherheit durch Barbiturat-Medikation und durch Kombination Barbiturat Alkohol. *Arzneimittel-Forsch* 12:1050.

Dossing, M., and Ranek, L. (1984). Isolated liver damage in chemical workers. *Br. J. Ind. Med.* 41:142–144.

Driver, H.E., and Swann, P.F. (1987). Alcohol and human cancer. *Anticancer Res.* 7:309–320.

Dunn, J.D., Clarkson, T.W., and Magos, L. (1978). Ethanol increased exhalation of mercury in mice. *Br. J. Ind. Med.* 35:241–244.

Early, J., McGrath, J., Messihn, F.S., and Hughes, M.J. (1983). Altered atrial responses to drugs after acute exposure to nitrogen dioxide and/or ethanol. *Drug and Chem. Toxicol.* 6:279–284.

Edwards, J.D., and Eckerman, D.A. (1979). Effects of diazepam and ethanol alone and in combination on conditioned suppression of keypecking in the pigeon. *Pharmacol. Biochem. Behav.* 10:217–221.

Eerola, R. (1961). The effect of ethanol on the toxicity of hexobarbital, thiopentone, morphine, atropine, and scapolamine. *Ann. Med. Exp. Biol. Fenniae* 39(3):7–70.

———. (1963). The effect of ethanol on the toxicity of promazine, chlorpromazine, promathazine and hydroxyzine. An experimental study on mice. *Acta Anaesthesiol. Scand.* 7:87–95.

Eerola, R., Venho, I., Vartiainen, O., and Venho, E.V. (1955). *Ann. Med. Exptl. Biol. Fenniae* 33:253.

Elbel, H. (1938). Neues zur blutalkoholfrage (Widmark-oder Friedmann-Klaas-Methode; Alkoholbestimmung in faulem blut; wirkung von aspirin, pyramidon, veronal und dextroenergen auf die blutalkoholkurue und auf die trunkenheit). *Disch. Z. Ges. Geriehtl. Med.* 30:218.

Elovaara, E., Collan, Y., Pfaffi, P., and Vainio, H. (1980). The combined toxicity of technical grade xylene and ethanol in the rat. *Xenobiotica* 10:435–445.

Emby, D.J., and Fraser, B.N. (1977). Hepatotoxicity of paracetamol enhanced by ingestion of alcohol: report of two cases. *S. Afr. Med. J.* 51:208.

Engstrom, K., Harkonen, H., Kalliokoski, P., and Rantanen, J. (1976). Urinary mandelic acid concentration after occupational exposure to styrene and its use as a biological exposure test. *Scand. J. Work Environ. Health* 2:21–26.

Enstrom, J.E. (1977). Colorectal cancer and beer drinking. *Br. J. Cancer* 35:674–683.

Evans, M.A., and Harbison, R.D. (1978). Cocaine-induced hepatotoxicity in mice. *Toxicol. Appl. Pharmacol.* 45:739–754.

Farinati, F., Zhou, Z., Bellah, J., Liever, C.S., and Garro, A.J. (1985). Effect of chronic ethanol consumption on activation of nitrosopyrrolidine to a mutagen by rat upper alimentary tract, lung, and hepatic tissue. *Drug Metab. Dispos.* 13:210–214.

Farinati, S., Espina, N., Lieber, C.S., and Garro, A.J. (1984). Ethanol-mediated inhibition of DNA repair: a specific effect on liver O_6 methylguanine in ethanol transferase. *Gastroenterology* 86(Part 2):1317.

Ferguson, R.K., and Vernon, R.J. (1970). Trichloroethylene in combination with CNS drugs. Effects of visual-motor tests. *Arch. Environ. Health* 20:462–467.

Fern, H.J., and Hodges, J.R. (1953). Synergic effects of amylobarbitone sodium and ethanol. *J. Pharm. Pharmacol.* 5:1041–1044.

Fischer, R.S., Walter, J.T., and Plummer, C.W. (1948). Quantitative estimation of barbiturates in blood by ultra-violet spectrophotometry. II. Experimental and clinical results. *Amer. J. Clin. Path.* 18:462.

Fleischmann, R., Scholte, W., Schomerous, H., Wolburg, H., Castrillon-Oberndorfer, W.L., and Hoensch, H. (1977). Kleinnotige leberzirrhose mit ausgepragter portaler hypertension als folge einer vitamin-A-intoxikation bei psoriasisbehandlung. *Deutsche Med. Wschr.* 102:1637–1640.

Folland, D.S., Schaffner, W., Grinn, H.E., Crofford, Q.B., and McMurray, D.R. (1976). Carbon tetrachloride toxicity potentiated by isopropyl alcohol: investigation of an industrial outbreak. *J. Amer. Med. Assoc.* 236:1853–1856.

Forney, R.B., Hulpieu, H.R., and Hughes, F.W. (1962). The comparative enhancement of the depressant action of alcohol by eight representative ataractic and analgesic drugs. *Experientia* 18:468–470.

Freund, G. (1979). Possible relationship of alcohol in membranes to cancer. *Cancer Res.* 39:2899–2901.

Fühner, H. (1930). Slg. Vergiftungsfalle A. 173 (Cited in Eerola, R. (1961). The effect of ethanol on the toxicity of hexabital, thiopental, morphine, atropine and scapolamine. *Ann. Med. Exp. Biol. Fenniae* 3a(3):7–70.

Gander, R.E. (1979). Psychoactive drug quantification by visual flicker sensitivity measurement. Thesis for the degree of Doctor of Philosophy, Institute of Biomedical Engineering, Department of Electrical Engineering, University of Toronto.

Ganey, P.E., and Thurman, R.G. (1988). Hepatotoxicity due to menadione is potentiated by ethanol in perfused rat liver. *Toxicologist* 8(1):217 (abst. 863).

Garcia de Torres, G., Romer, K.G., Torres Alanis, O., and Freundt, K.J. (1983). Blood acetaldehyde levels in alcohol-dosed rats after treatment with ANIT, ANTU, dithiocarbamate derivatives or cyanamide. *Drug and Chem. Toxicol.* 6(4):317–328.

Garner, R.C., and McLean, A.E.M. (1969). Increased susceptibility to carbon tetrachloride poisoning in the rat after pretreatment with oral phenobarbitone. *Biochem. Pharmacol.* 18:645–650.

Garro, J.A., Seitz, H.K., and Lieber, C.S. (1981). Enhancement of dimethylnitrosamine metabolism and activation to a mutagen following chronic ethanol consumption. *Cancer Res.* 41:120–124.

Gessner, P.K., and Cabana, B.E. (1964). The effect of ethanol on chloral hydrate hypnosis in mice. *Fed. Proc.* 23:348.

———. (1967). Chloral alcoholate: reevaluation of its role in the interaction between the hypnotic effects of chloral hydrate and ethanol. *J. Pharmacol. Exp. Therap.* 156:602–605.

———. (1970). A study of the interaction of the hypnotic effects and of the toxic effects of chloral hydrate and ethanol. *J. Pharmacol. Exp. Therap.* 174:247–259.

Gibel, W. (1969). Experimentelle Untersuchungen zur Synkarzinogenese beim Osophaguskarzinom. *Arch. Geschwulst.* 30:181–189.

Giles, H.G., Kadar, G., MacLeod, S.M., and Sellers, E.M. (1979). Use of the overturn endpoint in goldfish to measure the interaction of diazepam and thenaol and the effects of the binding of diazepam to bovine serum albumin. *Can. J. Physiol. Pharmacol.* 57:1186.

Gill, D.P., Kempen, R.R., Nash, J.B., and Ellis, S. (1979). Modifications of benzene myelotoxicity and metabolism by phenobarbital, SKF-525A, and 3-methylcholanthrene. *Life Sci.* 25:1633.

Glatt, H., deBalle, L., and Oesch, F. (1981). Ethanol or acetone pretreatment of mice strongly enhanced the bacterial mutagenicity of dimethylnitrosamine in assays mediated by liver subcellular fractions but not in host-mediated assays. *Carcinogenesis* 2:1057–1061.

Glinsukon, T., Taycharpipranai, S., and Toskulkao, C. (1978). Aflatoxin B_1 hepatotoxicity in rats pretreated with ethanol. *Experimentia* 34:869–870.

Goldfinger, R., Ahmed, K.S., Pitchumoni, C.S., et al. (1978). Concomitant alcohol and drug abuse enhancing acetaminophen toxicity. *Am. J. Gastroenterol.* 70:385–388.

Goodman, L., and Gilman, A. (1941). *The Pharmacological Basis of Therapeutics.* Macmillan, New York.

Gopinath, C., and Ford, E.J.H. (1975). The role of microsomal hydroxylases in the modification of chloroform hepatotoxicity in rats. *Br. J. Exp. Pathol.* 56:412–422.

Goulston, K., and Cooke, A.R. (1968). Alcohol, aspirin and gastrointestinal bleeding. *Br. Med. J.* 4:664–665.

Graham, J.D.P. (1960). Ethanol and the absorption of barbiturate. *Toxicol. Appl. Pharmacol.* 2:14–22.

Greenblatt, D.J., Shader, R.I., Weinberger, D.R., Allen, M.D., and MacLaughlin, D.S. (1978). Effect of a cocktail on diazepam absorption. *Psychopharmacology (Berlin)* 57:199–203.

Griciute, L., Castegnaro, M., and Bereziat, J.C. (1981). Influence of ethyl alcohol on carcinogenesis with N-nitrosodimethylamine. *Cancer Lett.* 13:345–352.

Gruber, C.M. (1955). A theoretical consideration of additive and potentiated effects between drugs with a practical example using alcohol and barbiturates. *Arch. Intern. Pharmacodyn.* 102:17–32.

Guild, W.R., Young, J.V., and Merrill, J.P. (1958). Anuria due to carbon tetrachloride. *Ann. Intern. Med.* 48:1221.

Gupta, R.C., and Kofoed, J. (1966). Toxicological statistics for barbiturates, other sedatives and tranquilizers in Ontario. *Can. Med. Assoc. J.* 94:863.

Habs, M., and Schmahl, D. (1981). Inhibition of the hepatocarcinogenic activity of diethylnitrosamine (DENA) by ethanol in rats. *Acta Gastroenterol.* 28:242–244.

Haggard, H.W., Greenberg, L.A., Rakieten, N., and Cohen, L.H. (1940). Studies on absorption, distribution and elimination of alcohol; influence of inhalation of oxygen and carbon dioxide and certain drugs on concentration of alcohol in blood causing respiratory failure. *J. Pharmacol. Exp. Therap.* 69:266–327.

Hall, M.C. (1921). Carbon tetrachloride for the removal of parasitic worms, especially hookworms. *J. Agri. Res.* 21(2):157–175.

Hanasono, G.K., Cote, M.G., and Plaa, G.L. (1975a). Potentiation of carbon tetrachloride-induced hepatotoxicity in alloxan- or streptozotocin-diabetic rats. *J. Pharmacol. Exp. Therap.* 192:592–604.

Hanasono, G.K., Witschi, H., and Plaa, G.L., (1975b). Potentiation of the hepatotoxic responses to chemicals in alloxan-diabetic rats. *Proc. Soc. Exp. Biol. Med.* 149:903–907.

Harris, R.N., and Anders, M.W. (1980). Effect of fasting, diethyl maleate, and alcohols on carbon tetrachloride–induced hepatotoxicity. *Toxicol. Appl. Pharmacol.* 56:191–198.

———. (1981). Phosgene: a possible role in the potentiation of carbon tetrachloride hepatotoxicity by 2-propanol. *Life Sci.* 29:503–507.

Harris, R.N., Ratnayake, J.H., Garry, V.F., and Anders, M.W. (1982). Interactive hepatotoxicity of chloroform and carbon tetrachloride. *Toxicol. Appl. Pharmacol.* 63:281–291.

Hasumura, Y., Teschke, R., and Lieber, C.S. (1974). Increased carbon tetrachloride hepatotoxicity and its mechanism after chronic ethanol consumption. *Gastroenterology* 66:415–422.

Hayes, S.L., Pablo, G., Randomski, K.T., et al. (1977). Ethanol and oral diazepam absorption. *New Eng. J. Med.* 296:186–189.

Heck, K., Mallach, H.J., Mayer, B., and Mayer, K. (1966). Uber die gemeinsame Wirkung von Alkohol and DMSO beim Menschen. *Med. Welt.* 17:963–975.

Hefner, R.E., Jr., Watanabe, P.G., and Gehring, P.J. (1975). Preliminary studies of the fate of inhaled vinyl chloride monomer (VCM) in rats. *Ann. NY Acad. Sci.* 246:135.

Herlong, H.F., Russell, R.M., and Maddrey, W.C. (1981). Vitamin A and zinc therapy in primary biliary cirrhosis. *Hepatology* 1:348–351.

Herr, F., Stewart, J., and Charest, M.P. (1961). Tranquilizers and antidepressants: a pharmacological comparison. *Arch. Intern. Pharmacodyn.* 134:328–342.

Hewitt, L.A., Ayotte, P., and Plaa, G.L. (1985). Modifications in rat hepatobiliary function following treatment with ketones and chloroform. *Toxicologist* 5:159.

Hewitt, L.A., Hewitt, W.R., and Plaa, G.L. (1983a). Fractional hepatic localization of $^{14}CHCl_3$ in mice and rats treated with chlordecone or mirex. *Fund. Appl. Toxicol.* 3:489–495.

Hewitt, L.A., Valiquette, C., and Plaa, G.L. (1983b). Correlation of biotransformation-detoxication parameters with 2-hexanone (MBK), 2-butanone (MEK) and acetone-potentiated chloroform ($CHCl_3$) hepatotoxicity. *Toxicologist* 3:99.

Hewitt, W.R., and Brown, E.M. (1984). Nephrotoxic interactions between ketonic solvents and halogenated aliphatic chemicals. *Fund. Appl. Toxicol.* 4:902–908.

Hewitt, W.R., Brown, E.M., and Plaa, G.L. (1983a). Relationship between the carbon skeleton length of solvents and potentiation of chloroform-induced hepatotoxicity in rats. *Toxicol. Lett.* 16:297–304.

———. (1983b). Acetone-induced potentiation of trihalomethane toxicity in male rats. *Toxicol. Lett.* 16:285–296.

Hewitt, W.R., Miyajima, H., Cote, M.G., and Plaa, G.L. (1979). Acute alteration of chloroform-induced hepato- and nephrotoxicity by mirex and Kepone. *Toxicol. Appl. Pharmacol.* 48:509–527.

———. (1980a). Modification of haloalkane-induced hepatotoxicity by exogenous ketones and metabolic ketosis. *Fed. Proc.* 39:3118–3123.

———. (1980b). Acute alteration of chloroform-induced hepato- and nephrotoxicity by n-hexane, methyl n-butyl ketone, and 2,5-hexanedione. *Toxicol. Appl. Pharmacol.* 53:230–248.

Hewitt, W.R., and Plaa, G.L. (1979). Potentiation of carbon tetrachloride–induced hepatotoxicity by 1,3-butanediol. *Toxicol. Appl. Pharmacol.* 47:177–180.

———. (1983). Dose-dependent modification of 1,1-dichloroethylene toxicity by acetone. *Toxicol. Lett.* 16:145–152.

Higgins, J.A., and McGuigan, H.A. (1983). The influence of caffeine on the effect of acetanilid. *J. Pharmacol. Exp. Therap.* 49:466.

Hillbom, M.E., Sarviharju, M.S., and Lindros, K.O. (1983). Potentiation of ethanol toxicity by cyanamide in relation to acetaldehyde accumulation. *Toxicol. Appl. Pharmacol.* 70:133–139.

Hindmarch, I. (1978). The effects of repeated doses of temazepam taken in the morning following nighttime medication. *Arzneimittel-Forsch* 28:2357–2360.

Ho, B.T., Fritchie, G.E., Englert, L.F., McIsaac, W.M., and Indanpaan-Heikka, J.E. (1971). Marihuana: importance of the route of administration. *J. Pharm. Pharmacol.* 23:309–310.

Holten, C.H., and Larsen, V. (1956). Potentiating effect of benactyzine derivatives and some other compounds on evipal anaesthesia in mice. *Acta Pharmacol. Toxicol.* 12:346–363.

Hoyumpa, A., Patwardhan, R., Maples, M., Desmond, P., Johnson, R., and Schenker, S. (1980). Effect of short term ethanol administration on lorazepam metabolism. *Gastroenterology* 79:1027.

Hughes, F.W., Forney, R.B., and Richards, A.B. (1965). Comparative effect in human subjects of chlordiazepoxide, diazepam and placebo on mental and physical performance. *Clin. Pharmacol. Therap.* 6:139–145.

Hultmark, D., Sundh, K., Johansson, L., and Arrhenius, E. (1979). Ethanol inhibition of vinyl chloride metabolism in isolated rat hepatocytes. *Chem. Biol. Interact.* 25:1.

Hursh, J.B., Clarkson, T.W., Cherian, M.G., Vostal, J.J., and Vander Mallie, R. (1976). Clearance of mercury (Hg-197, Hg-203) vapor inhaled by human subjects. *Arch. Environ. Health* 31:302–309.

Hursh, J.B., Greenwood, M.R., Clarkson, T.W., Allen, J., and Demouth, S. (1980). The effect of ethanol on the fate of mercury vapor inhaled by man. *J. Pharmacol. Exp. Therap.* 214:520–527.

Ikeda, M., and Ohtsuji, H. (1971). Phenobarbital-induced protection against toxicity of toluene and benzene in the rat. *Toxicol. Appl. Pharmacol.* 20:30–43.

Ilett, K.F., Reid, W.D., Sipes, I.G., and Krishna, G. (1973). Chloroform toxicity in mice: Correlation of renal and hepatic necrosis with covalent binding of metabolites to tissue macromolecules. *Exp. Mol. Pathol.* 19:215–229.

Ioannides, C., and Parke, D.V. (1973). The effect of ethanol administration on drug oxidations and possible mechanism of ethanol–barbiturate interaction. *Biochem. Soc. Trans.* 1:716–720.

Jentschura, G. (1965). Discussion remark. In *Dimethylsulfoxyd-DMSO, Schering-Symposium 1965*, G. Laudahn and H.J. Schlosshauer, Eds. E. Blaschker Verlag, Berlin, West Germany.

Jernigan, J.D., and Harbison, R.D. (1982). Role of bio-transformation in the potentiation of halocarbon hepatotoxicity by 2,5-hexanedione. *J. Toxicol. Env. Health* 9:761–781.

Jernigan, J.D., Pounds, J.G., and Harbison, R.D. (1983). Potentiation of chlorinated hydrocarbon toxicity by 2,5-hexanedione in primary cultures of adult rat hepatocytes. *Fund. Appl. Toxicol.* 3:22–26.

Jetter, W.W., and McLean, R. (1943). Poisoning by the synergistic effect of phenobarbital and ethyl alcohol. *AMA Arch. Pathol.* 36:112–122.

Johnson, M.K. (1967). Metabolism of chloroethanol in the rat. *Biochem. Pharmacol.* 16:185–199.

Kaplan, H.L., Jain, N.C., Forney, R.B., and Richards, A.B. (1969). Chloral hydrate-ethanol interaction in the mouse and dog. *Toxicol. Appl. Pharmacol.* 14:127–137.

Karobath, M., Rogers, J., and Bloom, F.E. (1980). Benzodiazepine receptors remain unchanged after chronic ethanol administration. *Neuropharmacology* (1):125–128.

Kaufman, S.E., and Kaye, M.D. (1978). Induction of gastroesophageal reflux by alcohol. *Gut.* 18:336–338.

Kaysen, G.A., Pond, S.M., Roper, M.H., Menke, D.J., and Marrama, M.A. (1985). Combined hepatic and renal injury in alcoholics during therapeutic use of acetaminophen. *Arch. Intern. Med.* 145:2019–2023.

Kershaw, W.C., Iga, T., and Klaassen, C.D. (1988). Ethanol (ETH) decreases the toxicity of Cd. *Toxicologist* 8(1):221 (abst. 82).

Kesteloot, H., Roelandt, J., Willems, J., Claess, J.H.H., and Joossens, J.V. (1968). An inquiry into the role of cobalt in the heart disease of chronic beer drinkers. *Circulation* 37:854.

Ketcham, A.S., Wexler, H., and Mantel, N. (1963). Effects of alcohol in mouse neoplasia. *Cancer Res.* 23:667–670.

Khan, A.U., Forney, R.B., and Hughes, F.W. (1964). Effect of tranquilizers on the metabolism of ethanol. *Arch. Intern. Pharmacodyn.* 150:171–176.

Kissin, B., and Kaley, M.M. (1974). Alcohol and cancer. In *The Biology of Alcoholism,* Vol. 3, B. Kissin and H. Begleiter, Eds. Plenum Publishing, New York, pp. 481–511.

Klaassen, C.D., and Plaa, G.L. (1966). Relative effects of various chlorinated hydrocarbons on liver and kidney function in mice. *Toxicol. Appl. Pharmacol.* 9:139–151.

Kluwe, W.M., and Hook, J.B. (1977). Polybrominated biphenyl potentiation of acute $CHCl_3$ toxicity. *Pharmacologist* 19:199.

———. (1978a). Chemically induced modification of chlorinated hydrocarbon solvent nephrotoxicity. *Toxicol. Appl. Pharamcol.* 45:228.

———. (1978b). Polybrominated biphenyl-induced potentiation of chloroform toxicity. *Toxicol. Appl. Pharmacol.* 45:861–869.

———. (1978c). Organ specific chloroform toxicity in mice: relationships to xenobiotic metabolism, glutathione depletion and covalent binding to microsomal protein. *Pharmacologist* 20:259.

Kluwe, W.M., McCormack, K.M., and Hook, J.B. (1978a). Potentiation of hepatic and renal toxicity of various compounds by prior exposure to polybrominated biphenyls. *Environ. Health Perspec.* 23:241–246.

———. (1978b). Selective modification of the renal and hepatic toxicities of chloroform by induction of drug-metabolizing enzyme systems in kidney and liver. *J. Pharmacol. Exp. Therap.* 207:566–573.

Kochman, R.L., Hirsch, J.D., and Clay, G.A. (1981). *Res. Commun. Sub. Abuse* 2:135.

Koop, D.R., Crump, B.L., Nordblom, G.D., and Coon, M.J. (1985). Immunochemical evidence for induction of the alcohol-oxidizing cytochrome P-450 of rabbit microsomes by diverse agents: ethanol, imidazole, trichloroethylene, acetone, pyrazole and isoniazid. *Proc. Natl. Acad. Sci.* 82:4065–4069.

Kopf, R. (1957). Zur Frage der Alkoholpotenzierung durch Phenothiazine. *Arch. Intern. Pharmacodyn.* 110:56–64.

Kramer, A., Staudinger, H.J., and Ullrick, V. (1974). Effect of n-hexane inhalation on the mono-oxygenase system in mice liver microsomes. *Chem. Biol. Interact.* 8:11–18.

Krebs, C. (1928). Experimeteiler Alkoholkrebs bei weissen Mausen. *Z. Immunol. Exp. Ther.* 59:203–218.

Kuratsune, M., Kohchi, S., and Horie, A. (1965). Carcinogenesis in the esophagus. I. Penetration of benzo(a)pyrene and other hydrocarbons into the esophageal mucosa. *Gann.* 56:177–187.

Kutob, S.D., and Plaa, G.L. (1962). The effect of acute ethanol intoxication on chloroform induced liver damage. *J. Pharmacol. Exp. Therap.* 135:245–251.

Laisi, U., Linnoila, M., Seppala, T., et al. (1979). Pharmacokinetic and pharmacodynamic interactions of diazepam with different alcoholic beverages. *Eur. J. Clin. Pharmacol.* 16:263–270.

Lake, B.G., Heading, C.E., Phillips, J.C., Gangolli, S.D., and Lloyd, A.G. (1974). Some studies on the metabolism in vitro of dimethylnitrosamine by rat liver. *Biochem. Soc. Trans.* 2:610–612.

Lambert, S.M. (1922). Carbon tetrachloride in the treatment of hookworm disease. *JAMA* 79(25):2055–2057.

Lamson, P.D., Gardner, G.H., Gustafson, R.K., Maire, E.D., McLean, A.J., and Wells, H.S. (1924). The pharmacology and toxicology of carbon tetrachloride. *J. Pharmacol. Exp. Therap.* 22:215.

Lancet. (1970). Gastrointestinal bleeding: what progress? 1:1157–1158.

———. (1980). Beer and bowel cancer. 1:1396–1397.

Langman, M.J.S. (1970). Epidemiological evidence for the association of aspirin and acute gastrointestinal bleeding. *Gut.* 11:627–634.

Lavigne, J.G., and Marchand, C. (1974). The role of metabolism in chloroform hepatotoxicity. *Toxicol. Appl. Pharmacol.* 29:312–326.

Lawson, J.W.R., Guynn, R.W., Cornell, N., and Veech, R.L. (1976). A possible role for pyrophosphate in control of hepatic glucose metabolism. In *Gluconeogenesis: Its Regulation in Mammalian Species,* R.W. Hanson and M.A. Mehlman, Eds. John Wiley and Sons, New York, pp. 498–506.

Lefevre, A., Adler, H., and Lieber, C.S. (1970). Effect of ethanol on ketone metabolism. *J. Clin. Invest.* 49:1775–1782.

Leibman, K.C., and Ortiz, E. (1977). Metabolism of halogenated ethylenes. *Environ. Health Perspec.* 21:91–97.

Leo, M., Arai, M., Sato, M., and Lieber, C.S. (1982). Hepatotoxicity of vitamin A and ethanol in the rat. *Gastroenterology* 82:194–205.

Leo, M., and Lieber, C.S. (1983). Hepatic fibrosis after long-term administration of ethanol and moderate vitamin supplementation in the rat. *Hepatology* 3:1–11.

Lesser, P.B., Vietti, M.M., and Clark, W.D. (1986). Lethal enhancement of therapeutic doses of acetaminophen by alcohol. *Digest. Dis. Sci.* 31(1):103–105.

Lester, D., and Benson, G.D. (1970). Alcohol oxidation in rats inhibited by pyrazole, oximes and amides. *Science* 169:282–283.

Lieber, C.S. (1974). Contrast between the acute and chronic effects of ethanol on lipid, ketone and drug metabolism. In *Regulation of Hepatic Metabolism,* E. Lundquist and N. Tygstrup, Eds. Academic Press, New York, pp. 379–400.

———. (1985). Ethanol metabolism and pathophysiology of alcoholic liver diseases. In *Alcohol Related Diseases in Gastroenterology,* H.K. Seitz, and B. Kommerelli, Eds. Springer, Berlin, pp. 19–47.

Lieber, C.S., and DeCarli, L.M. (1970). Quantitative relationship between the amount of dietary fat and the severity of the alcoholic fatty liver. *Am. J. Clin. Nutr.* 23:474–478.

———. (1977). Metabolic effects of alcohol on the liver. In *Metabolic Aspects of Alcoholism,* C.S. Lieber, Ed. MTP Press, Lancaster, pp. 31–79.

Lieber, C.S., Seitz, K.H., Garro, A.J., and Worner, T.M. (1979). Alcohol-related diseases and carcinogenesis. *Cancer Res.* 39:2863–2886.

Liljequist, R., Linnoila, M., Mattila, M.J., et al. (1975). Effects of two weeks' treatment with thioridazine, chlorpromazine, sulpiride and bromazepam, alone or in combination with alcohol, on learning and memory in man. *Psychopharmacologia* 44:205–208.

Liljequist, R., Palva, E., and Linnoila, M. (1979). Effects on learning and memory of 2-week treatments with chlordiazepoxide lactam, N-desmethyl-diazepam, oxazepam and methyloxazepam, alone or in combination with alcohol. *Int. Pharmacopsychiatry* 14:190–198.

Linnoila, M., and Hakkinen, S. (1974). Effects of diazepam and codeine, alone and in combination with alcohol, on simulated driving. *Clin. Pharmacol. Therap.* 15:368–373.

Linnoila, M., and Mattila, M.J. (1973). Drug interaction on psychomotor skills related to driving, diazepam and alcohol. *Eur. J. Clin. Pharmacol.* 5:186–194.

Linnoila, M., Otterstrom, S., and Anttila, M. (1974a). Serum chlordiazepoxide, diazepam and thioridazine concentrations after the simultaneous ingestion of alcohol or placebo drink. *Ann. Clin. Res.* 6:4–6.

Linnoila, M., Saario, I., and Maki, M. (1974b). Effect of treatment with diazepam or lithium and alcohol on psychomotor skills related to driving. *Eur. J. Clin. Pharmacol.* 7:337–342.

Linnoila, M., Saario, I., Olkoniemi, J., et al. (1975). Effect of two weeks' treatment with chlordiazepoxide or flupenthioxole, alone or in combination with alcohol, on psychomotor skills related to driving. *Arzneimittel-Forsch* 25:1088–1092.

Lish, P.M., Albert, J.R., Peters, E.L., and Allen, L.E. (1960). Pharmacology of methdilazine (Tacaryl). *Arch. Intern. Pharmacodyn.* 129:77.

Lyle, W.H., Spence, W.M., McKinneley, W.M., and Duckers, K. (1979). Dimethylformamide and alcohol intolerance. *Br. J. Ind. Med.* 36:63–66.

MacDonald, J.R., Gandolfi, A.J., and Sipes, I.G. (1982). Acetone potentiation of 1,1,2-trichloroethane hepatotoxicity. *Toxicol. Lett.* 13:57–69.

Magos, L. Clarkson, T.W., and Greenwood, M.R. (1973). The depression of pulmonary retention of mercury vapor by ethanol: identification of the site of action. *Toxicol. Appl. Pharmacol.* 26:180–183.

Magos, L., Halbach, S., and Clarkson, T.W. (1978). Role of catalase in the oxidation of mercury vapor. *Biochem. Pharmacol.* 27:1373–1377.

Mahaffey, K.R., Goyer, R.A., and Wilson, M.H. (1974). Influence of ethanol ingestion on lead toxicity in rats fed isocaloric diets. *Arch. Environ. Health* 28:217–224.

Maier, H., Born, I.A., Veith, S., Adler, D., and Seitz, H.K. (1986). The effect of chronic ethanol consumption on salivary gland morphology and function in the rat. *Alcohol Clin. Exp. Res.* 10:425–427.

Maita, K., Nakashima, N., and Shirasu, Y. (1988). Modification of hepatotoxicity of TOCP and MO by chronic ethanol ingestion. *Toxicologist* 8(1):178 (abst. 710).

Makar, A.B., Tephly, T.R., and Mannering, G.J. (1968). Methanol metabolism in the monkey. *Mol. Pharmacol.* 4:471–483.

Maling, H.M., Eichelbaum, F.M., Saul, W., Sipes, I.G., Brown, E.A.B., and Gillette, J.R. (1974b). Nature of the protection against carbon tetrachloride–induced hepatotoxicity produced by pretreatment with dibenamine [N-(2-chloroethyl) dibenzylamine]. *Biochem. Pharmacol.* 23:1479–1491.

Maling, H.M., Highman, B., Williams, M.A., Saul, W., Butler, W.M., Jr., and Brodies, B.B. (1974a). Reduction by pretreatment with dibenamine of hepatotoxicity induced by carbon tetrachloride, thioacetamide or dimethylnitrosamine. *Toxicol. Appl. Pharmacol.* 27:380–394.

Maling, H.M., Stripp, B., Sipes, I.G., Highman, B., Saul, W., and Williams, M.A. (1975). Enhanced hepatotoxicity of carbon tetrachloride, thioacetamide and dimethylnitrosamine by pretreatment of rats with ethanol and some comparisons with potentiation by isopropanol. *Toxicol. Appl. Pharmacol.* 33:291–308.

Mallach, H.J. (1967). Interaction of DMSO and alcohol. *Ann. NY Acad. Sci.* 141:457–462.

Mallach, H.J., and Roseler, P. (1961). Observations and research on the combined effects of alcohol and carbon monoxide. *Arzneimittel-Forsch* 11:10004–10008.

Mallov, S., and Baesl, T.J. (1972). Effect of ethanol on rates of elimination and metabolism of zoxazolamine, hexobarbital and warfarin sodium in the rat. *Biochem. Pharmacol.* 21:1667–1678.

Maltoni, C., and Lefemine, G. (1974). Carcinogenicity bioassays of vinyl chloride. I. Research plans and early results. *Environ. Res.* 7:387.

Mansuy, D., Beaune, P., Cresteil, T., Lange, M., and Leroux, J.P. (1977). Evidence for phosgene formation during liver microsomal oxidation of chloroform. *Biochem. Biophys. Res. Commun.* 79:513–517.

Maynert, E.W. (1965). Sedatives and hypnotics. In *Pharmacology in Medicine,* Third ed., J.R. Di Palma, Ed. McGraw-Hill Book Company, New York, pp. 161–171.

McCoy, G.D., Chen, C.B., and Hecht, S.S. (1979). Enhanced metabolism and mutagenesis of nitrosopyrrolidine in liver fractions isolated from chronic ethanol-consuming hamsters. *Cancer Res.* 39:793–796.

———. (1980). Influence of modifers of MFO activity on the in vitro metabolism of cyclic nitrosamines. In *Microsomes, Drug Oxidations and Chemical Carcinogenesis,* Vol. 2, M.J. Coon, A.H. Conney, R.W. Estabrook, H.V. Gelboin, S.R. Gillette, and P.J. O'Brien, Eds. Academic Press, New York, pp. 1189–1192.

McCoy, G.D., Hecht, S.S., Katayama, S., and Wynder, E.L. (1981). Differential effect of chronic ethanol consumption on the carcinogenicity of N-nitrosopyrrolidine and N^1-nitrosonornicotine in male Syrian Golden hamsters. *Cancer Res.* 41:2849–2854.

McGee, L.C., McCausland, A., Plume, C.A., and Marlett, N.C. (1942). Metabolic disturbances in workers exposed to dinitrotoluene. *J. Dig. Dis.* 39:329–332.

McLain, C.J., Kromhout, J.P., Peterson, F.J., and Holtzman, J.L. (1980). Potentiation of acetaminophen hepatotoxicity by alcohol. *JAMA* 244:251.

McLean, A.E.M., and McLean, E.K. (1966). The effect of diet of 1,1,1-trichloro-2,2-bis-(p-chlorophenyl) ethane (DDT) on microsomal hydroxylating enzymes and on sensitivity of rats to carbon tetrachloride poisoning. *Biochem. J.* 100:564–571.

McMichael, A.J., Potter, J.D., and Hetzel, B.S. (1979). Time trends in colorectal cancer mortality in relation to food and alcohol consumption: United States, United Kingdom, Australia and New Zealand. *Int. J. Epidemiol.* 8:295–303.

Mehar, G.S., Parker, J.M., and Tubas, T. (1974). Interaction between alcohol, minor tranquilizers and morphine. *Int. J. Clin. Pharmacol.* 9:70–74.

Melville, K.I., Joron, G.E., and Douglas, D. (1966). Toxic and depressant effects of alcohol given orally in combination with glutethimide or secobarbital. *Toxicol. Appl. Pharmacol.* 9:363–375.

Meredith, T.J., Vale, J.A., and Goulding, R. (1981). The epidemiology of acetaminophen poisoning in England and Wales. *Arch. Intern. Med.* 141(3):397–400.

Mezey, E. (1976). Ethanol metabolism and ethanol drug interactions. *Biochem. Pharmacol.* 25:869–875.

Miller, K.W., and Yang, C.S. (1984). Studies on the mechanisms of induction of n-nitrosodimethylamine demethylase by fasting, acetone, and ethanol. *Arch. Biochem. Biophys.* 229:483–491.

Miller, M.L., Radike, M.J., Andringa, A., and Bingham, E. (1982). Mitochondrial changes in hepatocytes of rats chemically exposed to vinyl chloride and ethanol. *Environ. Res.* 29:272–279.

Milner, T.G. (1970). Interaction between barbiturates, alcohol and some psychotropic drugs. *Med. J. Australia* 1:1204–1207.

Milner, G., and Landauer, A.A. (1973). Haloperidol and diazepam alone and together with alcohol in relation to driving safety. *Blutalkohol* 10:247–254.

Morgan, E.T., Koop, D.R., and Coon, M.J. (1983). Comparison of six rabbit liver cytochrome P-450 isozymes in formation of a reactive metabolite of acetaminophen. *Biochem. Biophys. Res. Commun.* 112:8–13.

Morgan, J.M., Hartley, M.W., and Miller, R.E. (1966). Nephropathy in chronic lead poisoning. *Arch. Intern. Med.* 118:17–29.

Morland, J., Setekleiv, J., Haffner, J.F.W., et al. (1974). Combined effects of diazepam and ethanol on mental and psychomotor functions. *Acta Pharmacol. Toxicol.* 34:5–15.

Muller, G., Spassowski, M., and Henschler, D. (1975). Metabolism of trichloroethylene in man. III. Interaction of trichloroethylene and ethanol. *Arch. Toxicol.* 33:173–189.

Murray, H.S., Strottman, M.P., and Cooke, A.R. (1974). Effect of several drugs on gastric potential differences in man. *Br. Med. J.* 1:19–21.

Nakajima, T., Okuyama, S., Yonekura, I., and Sato, A. (1985). Effects of ethanol and phenobarbital administration on the metabolism and toxicity of benzene. *Chem. Biol. Interact.* 55:23–38.

Needham, C.D., Kyle, J., Jones, P.F., Johnston, S.J., and Kerridge, D.F. (1971). Aspirin and alcohol in gastrointestinal hemorrhage. *Gut.* 12:819–821.

Nelson, R.L. (1984). Is the changing pattern of colorectal cancer caused by selenium deficiency? *Dis. Colon Rectum* 27:459–461.

Nelson, R.L., and Samelson, S.L. (1985). Neither dietary ethanol nor beer augments experimental colon carcinogenesis in rats. *Dis. Colon Rectum* 28:460–462.

Nicholson, A.N. (1979). Effect of alcohol on performance after nordiazepam. In *Advances in Pharmacology and Therapeutics,* Proceedings of the Seventh International Congress of Pharmacology, Paris, 1978, Vol. 8, G. Olive, Ed. Pergamon Press, Oxford, pp. 251–262.

Nielsen-Kudsk, F. (1965). The influence of ethyl alcohol on the absorption of mercury vapour from the lungs of man. *Acta Pharmacol. Toxicol.* 23:263–274.

———. (1969). Uptake of mercury vapour in blood in vivo and in vitro from Hg-containing air. *Acta Pharmacol. Toxicol.* 27:149–160.

Norppa, H., Sorsa, M., and Vaino, H. (1980). Chromosomal aberrations in bone marrow of Chinese hamsters exposed to styrene and ethanol. *Tox. Lett.* 5:241–244.

Nutr. Rev. (1983). Hypervitaminosis A enhances hepatotoxicity of ethanol in rats. 41(7):224–226.

Obidoa, O., and Okolo, T.C. (1979). Effect of ethanol administration on the metabolism of aflatoxin B1. *Biochem. Med.* 22:145–148.

O'Dell, J.R., Zetterman, R.K., and Burnett, D.A. (1986). Centrilobular hepatic fibrosis following acetaminophen-induced hepatic necrosis in an alcoholic. *JAMA* 255:2636.

Olszycka, L. (1935). Etude quantitative des phenomenes de synergie. Potentialisation de l'action hypnotique chez la souris. *Compt. Rend.* 201:796–797.

———. (1936). Etude quantitative des phenomenes de synergie. Contribution a l'etude du mecanisme des phenomenes de potentialisation de l'action hypnotique chez le rat. *Compt. Rend.* 202:1107–1109.

Opacka, J., Baranski, B., and Wronska-Nofer, T. (1985a). Effect of alcohol intake on some disturbances induced by chronic exposure to carbon disulfide in rats. I. Behavioural alterations. *Toxicol. Lett.* 23:91–97.

Opacka, J., Wronska-Nofer, T., Kolakowski, J., and Opalska, B. (1985b). Effect of alcohol intake on some disturbances induced by chronic exposure to carbon disulfide in rats. II. Biochemical and ultrastructural alterations in the peripheral nerves. *Toxicol. Lett.* 24:171–177.

Orten, J.M. (1935). On the mechanism of the hematopoietic action of cobalt. *Am. J. Physiol.* 114:414.

Palva, E.S., and Linnoila, M. (1978). Effect of active metabolites of chlordiazepoxide and diazepam, alone or in combination with alcohol, on psychomotor skills related to driving. *Eur. J. Clin. Pharmacol.* 13:345–350.

Pantarotto, C., Salmona, M., Szczawinska, K., and Bidoli, F. (1980). Gas chromatographic–mass spectrometric studies on styrene toxicity. In *Analytical Techniques in Environmental Chemistry*, J. A'lbaiges, Ed. Pergamon, Oxford, pp. 245–279.

Parboosingh, I.S. (1960). Chloral hydrate and rum analgesia in labor. *J. Int. Coll. Surg.* 34:769–773.

Parker, G.A., Bogo, V., and Young, R.W. (1981). Acute toxicity of conventional versus shale-derived JP-5 jet fuel: light microscopic, hematologic and serum chemistry studies. *Toxicol. Appl. Pharmacol.* 57:302–317.

Patel, J.M., Harper, C., and Drew, R.T. (1978). The biotransformation of p-xylene to a toxic aldehyde. *Drug Metab. Dispos.* 6:368–374.

Paton, W.D.M. (1975). Pharmacology of marijuana. *Ann. Rev. Pharmacol.* 15:191–220.

Paul, C.J., and Whitehouse, L.W. (1977). Metabolic basis for the supra-additive effect of the ethanol–diazepam combination in mice. *Br. J. Pharmacol.* 60:83–90.

Pegg, A.E., and Perry, W. (1981). Alkylation of nucleic acids and metabolism of small doses of dimethyl-nitrosamine in the rat. *Cancer Res.* 41:3128–3132.

Peng, R., Yong Tu, Y., and Yang, C.S. (1982). The induction and competitive inhibition of a high affinity microsomal nitrosodimethylamine demethylase by ethanol. *Carcinogenesis* 3:1457–1461.

Pessayre, D., Cobert, B., Descatoire, V., Degott, C., Babany, G., Funck-Brentano, C., Delaforge, M., and Larrey, D. (1982). Hepatotoxicity of trichloroethylene–carbon tetrachloride mixtures in rats. *Gastroenterology* 83:761–772.

Peter, H. (1939). Alcohol und sedativa. *Dtsch. Z. Ges. Gerichtl. Med.* 31:113–154.

Peterson, D.I., Peterson, J.E., Hardinge, M.G., and Wacker, W.E.C. (1963). Experimental treatment of ethylene glycol poisoning. *JAMA* 186:955–957.

Peterson, D.I., Peterson, J.E., and Hardinge, M.G. (1968). Protection by ethanol against the toxic effects of monofluoroethanol and monochloroethanol. *J. Pharm. Pharmacol.* 20:465–468.

Phillips, J.C., Lake, B.G., Gangolli, S.D., Grasso, P., and Lloyd, A.G. (1977). Effects of parazole and 3-amino-1,2,4-triazole on the metabolism and toxicity of dimethylnitrosamine in the rat. *J. Nat. Can. Inst.* 58:629–633.

Pilcher, J.D. (1912). Alcohol and caffeine: a study of antagonism and synergism. *J. Pharmacol. Exp. Therap.* 3:267–298.

Pirola, R.C. (1977). *Drug metabolism and alcohol.* University Park Press, Baltimore.

Pitts, L.L., Bruner, R.H., D'Addario, A.P., and Uddin, D.E. (1983). Induction of renal lesions following oral dosing with hydrocarbon fuels. *Toxicologist* 3:70.

Plaa, G.L., and Ayotte, P. (1985). Taurolithocholate-induced intrahepatic cholestasis: Potentiation by methyl isobutyl ketone and methyl n-butyl ketone in rats. *Toxicol. Appl. Pharmacol.* 80:228-234.

Plaa, G.L., Caille, G., Vezina, M., Iijima, M., and Gote, M.G. (1987). Chloroform interaction with chlordecone and mirex: correlation between biochemical and histological indices of toxicity and quantitative tissue levels. *Fund. Appl. Toxicol.* 9:198-207.

Plaa, G.L., Hewitt, W.R., DuSouich, P., Caille, G., and Lock, S. (1982). Isopropanol and acetone potentiation of carbon tetrachloride–induced hepatotoxicity: single versus repetitive pretreatments in rats. *J. Toxicol. Environ. Health* 9:235-250.

Plaa, G.L., and Traiger, G.J. (1972). Mechanism of potentiation of CCl_4-induced hepatotoxicity. In *Pharmacology and the Future of Man,* Proceedings of the Fifth International Congress on Pharmacology. Basal, Karger. 2:100-113.

Poda, G.A. (1966). Hydrogen sulfide can be handled safely. *Arch. Environ. Health* 12:795-800.

Pohl, L.R. (1979). Biochemical toxicology of chloroform. *Rev. Biochem. Toxicol.* 1:79-107.

Pohl, L.R., Bhooshan, B., Whittaker, N.F, and Krishna, G. (1977). Phosgene: a metabolite of chloroform. *Biochem. Biophys. Res. Commun.* 79:684-691.

Pohl, L.R., George, J.W., Martin, J.L., and Krishna, G. (1979). Deuterium isotope effect in in vivo bioactivation of chloroform to phosgene. *Biochem. Pharmacol.* 28:561-563.

Pohl, L.R., Martin, J.L., and George, J.W. (1980). Mechanism of metabolic activation of chloroform by rat liver microsomes. *Biochem. Pharmacol.* 29:3271-3276.

Pollack, E.S., Nomura, A., Herlbrun, L.K., Stemmermann, G.N., and Green, S.B. (1984). Prospective study of alcohol consumption and cancer. *N. Eng. J. Med.* 310:617-621.

Powis, G. (1975). Effect of a single oral dose of methanol, ethanol and propan-2-ol on the hepatic microsomal metabolism of foreign compounds in the rat. *Biochem. J.* 148:269-277.

Prescott, L.F., and Critchley, J.A.J.H. (1983a). Drug interactions affecting analgesic toxicity. *Am. J. Med.* (November 13):113-116.

———. (1983b). Ethanol in paracetamol poisoning. *Lancet* 2:617.

Prescott, L.F., Wright, N., Roscoe, P., et al. (1971). Plasma-paracetamol half-life and hepatic necrosis in patients with paracetamol overdosage. *Lancet* 1:519.

Price, N.D. (1938). Alcoholic beri-beri. *Lancet* 1:83.

Radike, M.J., Stemmer, K.L., and Bingham, E. (1981). Effect of ethanol on vinyl chloride carcinogenesis. *Environ. Health Perspec.* 41:59-62.

Radike, M.J., Stemmer, K.L., Brown, P.G., Larson, E., and Bingham, E. (1977). Effect of ethanol and vinyl chloride on the induction of liver tumors: preliminary report. *Environ. Health Perspec.* 21:153-155.

Rajtar, G. (1977). The influence of chlorpromazine, diazepam and imipramine on the central action of ethanol. *Arch. Immunol. Therap. Exp.* 25:813-818.

Ramsey, H., and Haag, H.B. (1946). The synergism between the barbiturates and ethyl alcohol. *J. Pharmacol. Exp. Therap.* 88:313-322.

Rangno, R., Sitar, D.S., Ogilvie, R.I., et al. (1976). The pharmacokinetic interaction of ethanol with amobarbital and diazepam. *Clin. Res.* 24:652.

Rerup, C.C. (1970). Drugs producing diabetes through damage of the insulin secreting cells. *Pharmacol. Rev.* 22:485–518.

Riihimaki, V., Laine, A., Sarolainen, K., and Sippel, H. (1982b). Acute solvent-ethanol interactions with special reference to xylene. *Scand. J. Work Environ. Health* 8:77–79.

Riihimaki, V., Savolainen, K., Pfaffli, P., Pekari, K., Sippel, H.W., and Laine, A. (1982a). Metabolic interaction between m-xylene and ethanol. *Arch. Toxicol.* 49:253–263.

Ritchie, J.M. (1965). In *The Pharmacologic Basis of Therapeutics,* Third ed., L.S. Goodman and A. Gilman, Eds. Macmillan, New York.

Roden, S., Harvey, P., and Mitchard, M. (1977). The influence of alcohol on the persistent effects on human performance of the hypnotics mandrax and nitrazepam. *Int. J. Clin. Pharmacol.* 15:350–355.

Rosenthal, S.M. (1930). Some effects of alcohol upon the normal and damaged liver. *J. Pharm. Exp. Therap.* 38:291.

Rubin, E., Gang, H., Misra, P.S., and Lieber, C.S. (1970). Inhibition of drug metabolism by acute ethanol intoxication: a hepatic microsomal mechanism. *Am. J. Med.* 49:801–806.

Rubin, E., and Lieber, C.S. (1968). Hepatic microsomal enzymes in man and rats: induction and inhibition by ethanol. *Science* 162:690–691.

Rubin, R.J., Taffe, B., and Egner, P. (1984). Interaction of alcohols with carbon disulfide metabolism. *Toxicologist* 4:151 (abst. 602).

Russell, R.M., Morrison, S.A., Smith, F.R., Oaks, E.V., and Carney, E.A. (1978). Vitamin A reversal of abnormal dark adaptation in cirrhosis. *Ann. Intern. Med.* 88:622–626.

Saario, I. (1976). Psychomotor skills during subacute treatment with thioridazine and bromazepam and their combined effects with alcohol. *Ann. Clin. Res.* 8:117–123.

Saario, I., and Linnoila, M. (1976). Effect of subacute treatment with hypnotics, alone or in combination with alcohol, on psychomotor skills related to driving. *Acta Pharmacol. Toxicol.* 38:382–392.

Saario, I., Linnoila, M., and Maki, M. (1975). Interaction of drugs with alcohol on human psychomotor skills related to driving. Effect of sleep deprivation on two weeks' treatment with hypnotics. *J. Clin. Pharmacol.* 15:52–59.

Saikkonen, J. (1959). Cobalt as a producer of porphyrinuria and polycythemia. *J. Lab. Clin. Med.* 54:860.

Salo, J.A. (1983). Ethanol-induced mucosal injury in rabbit esophagus. *Scand. J. Gastroenterol.* 18:713–721.

Sandstead, H.H., Michelakis, A.M., and Temple, T.E. (1970). Lead intoxication: its effect on the renin aldosterone response to sodium deprivation. *Arch. Environ. Health* 20:356–363.

Sato, A., and Nakajima, T. (1985). Enhanced metabolism of volatile hydrocarbons in rat liver following food deprivation, restricted carbohydrate intake, and administration of ethanol, phenobarbital, polychlorinated biphenyl and 3-methylcholanthrene: a comparative study. *Xenobiotica* 15:67

Sato, A., Nakajima, T., and Koyama, Y. (1980). Effects of chronic ethanol consumption on hepatic metabolism of aromatic and chlorinated hydrocarbons in rats. *Br. J. Ind. Med.* 37:382–386.

———. (1981a). Effects of acute and chronic ethanol consumption on the in vivo and in vitro metabolism of some volatile hydrocarbons. In *Industrial and Environmental Xenobiotics,* M. Cikrt and G.L. Plaa, Eds. Springer-Verlag, New York.

———. (1981b). Dose-related effects of a single dose of ethanol on the metabolism in rat liver of some aromatic and chlorinated hydrocarbons. *Toxicol. Appl. Pharmacol.* 60:8–15.

———. (1983). Interaction between ethanol and carbohydrate on the metabolism in rat liver of aromatic and chlorinated hydrocarbons. *Toxicol. Appl. Pharmacol.* 68:242.

Sato, C., and Lieber, C.S. (1981). Mechanisms of the preventive effect of ethanol on acetaminophen-induced hepatotoxicity. *J. Pharmacol. Exp. Therap.* 218:811.

Sato, C., Matsuda, Y., and Lieber, C.S. (1981a). Increased hepatotoxicity of acetaminophen after chronic ethanol consumption in the rat. *Gastroenterology* 80:140–148.

Sato, C., Nakano, M., and Lieber, C.S. (1981b). Prevention of acetaminophen-induced hepatotoxicity by acute ethanol administration in the rat: comparison with carbon tetrachloride–induced hepatotoxicity. *J. Pharmacol. Exp. Therap.* 218:805–810.

Savary, M., Miller, G., and Roethlisberger, B.C. (1981). Spezielle Probleme des Endobarchyosophagus. In *Refluxtherapie,* K.L. Blum and J.R. Siewert, Eds. Springer, New York, pp. 437–461.

Savolainen, K. (1980). Combined effects of xylene and alcohol in the CNS. *Acta Pharmacol. Toxicol.* 46:366–372.

Savolainen, K., and Riihimaki, V. (1981). Xylene and alcohol involvement of the human equilibrium system. *Acta Pharmacol. Toxicol.* 49:447–451.

Savolainen, K., Riihimaki, V., Vaheri, E., and Linnoila, M. (1980). Effects of xylene and alcohol on vestibular and visual functions in man. *Scand. J. Work Environ. Health* 6:94–103.

Schmahl, D. (1976). Investigations on esophageal carcinogenicity by methylphenylnitrosamine and ethanol in rats. *Cancer Lett.* 1:215–218.

Scholler, K.L. (1970). Modification of the effects of chloroform on the rat liver. *Br. J. Anaesth.* 42:603–605.

Schwartz, M., Appel, K.E., Schrenk, and Kunz, W. (1980). Effect of ethanol on microsomal metabolism of dimethylnitrosamine. *J. Cancer Res. Clin. Oncol.* 97:133–240.

Schwartz, M., Wiesbeck, G., Hummel, J., and Kunz, W. (1982). Effect of ethanol on dimethylnitrosamine activation and DNA synthesis in rat liver. *Carcinogenesis* 3(9):1071–1075.

Seef, L.B., Cuccherin, B.A., Zimmerman, H.J., Adler, E., and Benjamin, J.B. (1986). Acetaminophen hepatotoxicity in alcoholics. *Ann. Int. Med.* 104:399–404.

Seidel, G. (1967). Verteilung von pentobarbital barbital und thiopental unter Athanol. *Narnyn-Schmiedebergs Arch. Pharmak. V. Exp. Path.* 257:221–229.

Seitz, H.K. (1985). Ethanol and carcinogenesis. In *Alcohol Related Diseases in Gastroenterology,* H.K. Seitz and B. Kommerell, Eds. Springer, Berlin, pp. 196–212.

Seitz, H.K., Garro, A.J., and Lieber, C.S. (1978). Effect of chronic ethanol ingestion on intestinal metabolism and mutagenicity of benzo(a)pyrene. *Biochem. Biophys. Res. Commun.* 85:1061–1066.

———. (1979a). Increased activation of procarcinogens after ethanol. *Fed. Proc.* 38:1404.

———. (1981a). Enhanced pulmonary and intestinal activation of procarcinogens and mutagens after chronic ethanol consumption in the rat. *Eur. J. Clin. Invest.* 11:33–38.

———. (1981b). Sex dependent effect of chronic ethanol consumption in rats on hepatic microsome mediated mutagenicity of benzo(a)pyrene. *Cancer Lett.* 13:97–102.

Seitz, H.K., Korsten, M.A., and Lieber, C.S. (1979b). Ethanol oxidation by intestinal microsomes: microsomal activity after chronic ethanol administration. *Life Sci.* 25:1443–1448.

Seitz, H.K., and Simanowski, U.A. (1986). Ethanol and carcinogenesis of the alimentary tract. *Alcoholism* 10(6):33–40.

Sellers, E.M., and Busto, U. (1982). Benzodiazepines and ethanol: assessment of the effects and consequences of psychotropic drug interactions. *J. Clin. Psychopharmacol.* 2:249–262.

Sellers, E.M., Giles, G.H., Greenblatt, D.J., and Naranjo, C.A. (1980a). Differential effects on benzodiazepine disposition by disulfiram and ethanol. *Arzneimittel-Forsch* 30:882–886.

Sellers, E.M., and Kalant, H. (1978). Pharmacotherapy of acute and chronic alcoholism and alcohol withdrawal syndrome. In *Principles of Psychopharmacology,* W.G. Clark and J. del Giudice, Eds. Academic Press, New York, pp. 721–740.

Sellers, E.M., Naranjo, C.A., Giles, H.G., Frecker, R.C., and Beching, M. (1980b). Mechanism of ethanol-diazepam pharmacokinetic interactions. *Clin. Pharmacol. Therap.* 28:638–645.

Shideman, F.E. (1954). Sedatives and hypnotics. In *Pharmacology in Medicine,* V.A. Drill, Ed. McGraw-Hill Book Company, New York, Chapter 13, pp. 9–10.

Shirazi, S.S., and Platz, C.E. (1978). Effect of alcohol on canine esophageal mucosa. *J. Surg. Res.* 25:373–379.

Sipes, I.G., Asghar, K., Docks, E., Boykins, E., and Krishna, G. (1973b). Covalent binding as a basis for biochemical mechanisms of toxicities: comparison between covalent binding and hepatotoxicity produced by CCl_4 and CCl_3. *Br. Fed. Proc. Fed. Amer. Soc. Exp. Biol.* 32:319 (abst.).

Sipes, I.G., Slocumb, M.L., and Holtzman, G. (1978). Stimulation of microsomal dimethylnitrosamine-N-demethylase by pretreatment of mice with acetone. *Chem. Biol. Interact.* 21:155–166.

Sipes, I.G., Stripp, B., Krishna, G., Maling, H.M., and Gillette J.R. (1973a). Enhanced hepatic microsomal activity by pretreatment of rats with acetone or isopropanol. *Proc. Soc. Exp. Biol. Med.* 142:237–240.

Slater, T.F. (1966). Necrogenic action of carbon tetrachloride in the rat. A speculative mechanism based on activation. *Nature* 209:36–40.

Smillie, W.G., and Pessoa, S.B. (1923). Treatment of hookworm disease with carbon tetrachloride. *Am. J. Hyg.* 3:35–43.

Smith, A.C., Freeman, R.W., and Harbison, R.D. (1981). Ethanol enhancement of cocaine-induced hepatotoxicity. *Biochem. Pharm.* 30:453–458.

Smith, J.H., and Hook, J.B. (1983). Mechanism of chloroform nephrotoxicity. II. In vitro evidence for renal metabolism of chloroform in mice. *Toxicol. Appl. Pharmacol.* 70:480–485.

Smith, J.W., and Loomis, T.A. (1951). The potentiating effect of alcohol on thiopental induced sleep. *Proc. Soc. Exp. Biol. Med.* 78:827–829.

Smith, M.E., Evans, R.L., Newman, E.J., and Newman, H.W. (1961). Psychotherapeutic agents and ethyl alcohol. *Quart. J. Alc. Stud.* 22:241–249.

Smyth, R.D., Martin, G.J., Moss, J.N., and Beck, H. (1967). The modification of various enzyme parameters in brain acetylcholinesterase metabolism by chronic ingestion of ethanol. *Exp. Med. Surg.* 25:1.

Sofia, R.D., Kubena, R.K., and Barry, H., III. (1971). Comparison of four vehicles for intraperitoneal administration of Δ^9-tetrahydrocannabinol. *J. Pharm. Pharmacol.* 23:889–891.

Sofia, R.D., Kubena, R.K., and Barry, H., III. (1974). Comparison among four vehicles and four routes for administering Δ^9-tetrahydrocannabinol. *J. Pharm. Sci.* 63:939–941.

Sollman, T. (1957). *A Manual of Pharmacology.* W.B. Saunders Company, Philadelphia.

Spiegelhalder, B., Eisenbrand, G., and Preussman, R. (1982). Urinary excretion of N-nitrosamines in rats and humans. In *IARC Scientific Publication 41.* Lyon, pp. 443–449.

Spiegelhalder, B., and Preussmann, R. (1985). In vivo nitrosation of amidopyrene in humans: use of ethanol effect for biological monitoring of N-nitrosodimethylamine in urine. *Carcinogenesis* 6:545–548.

Stenger, R.J., and Johnson, E.A. (1971). Further observations upon the effects of phenobarbital pretreatment on the hepatotoxicity of carbon tetrachloride. *Exp. Molec. Path.* 14:220–227.

Stenger, R.J., Miller, R.A., and Williamson, J.N. (1970). Effects of phenobarbital pretreatment on the hepatotoxicity of carbon tetrachloride. *Exp. Molec. Path.* 13:242–252.

Steup, D.R., and Forney, R.B., Sr. (1988). Ethanol inhibits hepatic metabolism of intravenously administered morphine in rats. *Toxicologist* 8(1):267 (abst. 1066).

Stewart, R.D., Torkelson, T.R., Hake, C.L., and Erley, D.S. (1960). Infrared analysis of carbon tetrachloride and ethanol in blood. *J. Lab. Clin. Med.* 56:148–156.

Stolman, A. (1967). Combined action of drugs with toxicological implications—Part I. *Prog. Chem. Toxicol.* 3:305–361.

———. (1969). Combined action of drugs with toxicological implications—Part II. *Prog. Chem. Toxicol.* 48:257–395.

Strongin, E.I., and Winsor, A.L. (1935). The antagonistic action of coffee and alcohol. *J. Abnorm. Psychol.* 30:301–313.

Strubelt, O., Obermeier, F., and Siegers, C.P. (1978a). The influence of ethanol pretreatment on the effects of nine hepatotoxic agents. *Acta Pharmacol. Toxicol.* 43:211–218.

Strubelt, O., Obermeier, F., Siegers, C.P., and Volpel, M. (1978b). Increased carbon tetrachloride hepatotoxicity after low level ethanol consumption. *Toxicol.* 10:261–270.

Swann, P.F. (1982). Metabolism of nitrosamines: observations on the effect of alcohol on nitrosamine metabolism and on human cancer. *Banbury Report* 12:53–68.

Swann, P.F., Coe, A.M., and Mace, R. (1984). Ethanol and dimethylnitrosamine and diethylnitrosamine metabolism in the rat. Possible relevance to the influence of ethanol on human cancer incidence. *Carcinogenesis* 5:1337–1343.

Synder, C.A., Baarson, K.A., Goldstein, B.D., and Albert, R.E. (1981). Ingestion of ethanol increases the hematotoxicity of inhaled benzene in C57BL mice. *Bull. Environ. Contam. Toxicol.* 27:75–180.

Synderwine, E., Kroll, R., and Rubin, R. (1988). The possible role of the ethanol-inducible isozyme of cytochrome P450 in the metabolism and distribution of carbon disulfide. *Toxicol. Appl. Pharmacol.* 93:11–21.

Taeuber, K., Badian, M., Brettel, H.F., et al. (1979). Kinetic and dynamic interaction of clobazam and alcohol. *Br. J. Clin. Pharmacol.* 7:91S–95S.

Teschke, R., Minzlaff, M., Oldiges, H., and Frenzel, H. (1983). Effect of chronic alcohol consumption on human cancer incidence due to dimethylnitrosamine administration. *J. Cancer Res. Clin. Oncol.* 106:58–64.

Ticku, M.K., and Davis, W.C. (1981). Evidence that ethanol and pentabarbital enhance [3H] diazepam binding at the benzodiazepine-GABA receptor-ionophore complex indirectly. *Eur. J. Pharmacol.* 71:521–522.

Tipton, D.L., Sutherland, V.C., Burbridge, T.N., and Simon, A. (1961). Effect of chlorpromazine on blood level of alcohol in rabbits. *Am. J. Physiol.* 200:1007–1010.

Tishler, S.L., and Goldman, P. (1970). Properties and reactions of salicyl-coenzyme A. *Biochem. Pharmacol.* 19:143–150.

Tomera, J.F., Skipper, P.L., Wishnok, J.S., Tannenbaum, S.R., and Brunengraber, H. (1984). Inhibition of N-nitrosodimethylamine metabolism by ethanol in the isolated perfused rat liver. *Carcinogenesis* 5:113–116.

Traiger, G.J., and Bruckner, J.B. (1976). The participation of 2-butanone in 2-butanol-induced potentiation of carbon tetrachloride hepatotoxicity. *J. Pharmacol. Exp. Therap.* 196:493–500.

Traiger, G.J., and Plaa, G.L. (1971). Differences in the potentiation of carbon tetrachloride in rats by ethanol and isopropanol pretreatment. *Toxicol. Appl. Pharmacol.* 20:105–112.

———. (1972a). Relationship of alcohol metabolism to the potentiation of CCl_4 hepatotoxicity induced by aliphatic alcohols. *J. Pharmacol. Exp. Therap.* 183:481–488.

———. (1972b). Effect of pyrazole on the isopropanol-induced potentiation of CCl_4 hepatotoxicity. *Fed. Proc.* 31:519.

———. (1974). Chlorinated hydrocarbon toxicity: potentiation by isopropyl alcohol and acetone. *Arch. Environ. Health* 28:276–278.

Tredger, J.M., Smith, H.A., Read, R.B., Portmann, B., and Williams, R. (1985). Effects of ethanol ingestion on the hepatotoxicity and metabolism of paracetamol in mice. *Toxicol.* 36:341–352.

Tredger, J.M., Smith, H.A., Read, R.B., and Williams, R. (1986). Effects of ethanol ingestion on the metabolism of a hepatotoxic dose of paracetanol in mice. *Xenobiotica* 16(7):661–670.

Tu, Y.Y., Peng, R., Chang, Z.F., and Yang, C.S. (1983). Induction of a high affinity nitrosamine demethylase in rat liver microsomes by acetone and isopropanol. *Chem. Biol. Interact.* 44:247–260.

Tu, Y.Y., and Yang, C.S. (1983). High-affinity nitrosamine dealkylase system in rat liver microsomes and its induction by fasting. *Cancer Res.* 43:623–629.

Utesch, R.C., Weir, F.W., and Bruckner, J.V. (1981). Development of an animal model of solvent abuse for use in evaluation of extreme TCE inhalation. *Toxicol.* 19:169–182.

Veldstra, H. (1950). Synergism and potentiation. *Pharm. Rev.* 8:339–387.

Venkatesan, N., Argus, M.F., and Arcos, J.C. (1970). Mechanism of 3-methylcholanthrene-induced inhibition of dimethylnitrosamine demethylase in rat liver. *Cancer Res.* 30:2556–2562.

Vernot, E.H., and Pollard, D.L. (1983). Urinary metabolities of JP-4 jet fuel. *Toxicologist* 3:51.

Vezina, M., Ayotte, P., and Plaa, G.L. (1985). Potentiation of necrogenic and cholestatic liver injury by 4-methyl-2-pentanone. *Can. Fed. Biol. Soc.* 28:221.

Viola, P.L., Bigotti, A., and Caputo, A. (1971). Oncogenic response of rat skin, lungs and bones to vinyl chloride. *Cancer Res.* 31:516.

Volans, G.N. (1976). Self-poisoning and suicide due to paracetamol. *J. Int. Med. Res.* 4(4 suppl.):7–13.

Von Mallach, H.J., Moosmayer, A., Gottwald, K., et al. (1975). Pharmacokinetishce Untersuchungen uber Resorption und Ausscheidung von Oxazepam in Kombination mit Alkohol. *Arzneimittel-Forsch* 25:1840–1845.

Von Oettingen, W.F. (1955). The halogenated hydrocarbons, toxicity and potential danger. USPHS Publ. 414, U.S. Government Printing Office.

———. (1964). *The Halogenated Hydrocarbons of Industrial and Toxicological Importance.* Elsevier, Amsterdam.

Von Staak, M., and Moosmayer, A. (1978). Pharmacokinetic studies on interactions between dipotassium chlorazepate and alcohol after oral administration. *Arzneimittel-Forsch* 28:1187–1191

Wacker, W.E., Haynes, H., Druyan, R., Fisher, W. and Coleman, J.E. (1965). Treatment of ethylene glycol poisoning with ethyl alcohol. *JAMA* 194:173–175.

Waldron, H.A., Cherry, N., and Johnston, J.D. (1983). The effects of ethanol on blood toluene concentrations. *Int. Arch. Occup. Environ. Health* 51:365–369.

Walker, R.M., McElligott, T.F., Power, E.M., Massey, T.E., and Racz, W.B. (1983). Increased acetaminophen-induced hepatotoxicity after chronic ethanol consumption in mice. *Toxicology* 28:193.

Wallen, M., Naslund, P.H., and Byfalt Nordqvist, M. (1984). The effects of ethanol on the kinetics of toluene in man. *Toxicol. Appl. Pharmacol.* 76:414–419.

Watrous, W.M., and Plaa, G.L. (1971). The potentiation of $CHCl_3$-induced nephrotoxicity by some aliphatic alcohols in mice. *Pharmacologist* 13:227.

Webb, M. (1968). The biological action of cobalt and other metals. III. Chelation of cations by dihydrolipoic acid. *Biochem. Biophys. Acta* 65:47.

White, J.F., and Carlson, G.P. (1981). Epinephrine-induced cardiac arrhythmias in rabbits exposed to TCE potentiation by ethanol. *Toxicol. Appl. Pharmacol.* 80:460–471.

Whittlesey, P. (1954). The effects of pentobarbital on the metabolism of ethyl alcohol in dogs. *Bull. Johns Hopkins Hosp.* 95:81–88.

Wiberg, G.S., Coldwell, B.B., and Trenholm, H.L. (1969). Toxicity of ethanol–barbiturate mixtures. *J. Pharm. Pharmacol.* 21:232–236.

Wilson, H.K., Cocker, J., Purnell, C.J., Brown, R.H., and Gompertz, D. (1979). The time course of mandelic and phenylglyoxylic acid excretion in workers exposed to styrene under model conditions. *Br. J. Ind. Med.* 36:235–237.

Wilson, H.K., Robertson, S.M., Waldron, H.A., and Gompertz, D. (1983). Effect of alcohol on the kinetics of mandelic acid excretion in volunteers exposed to styrene vapour. *Br. J. Ind. Med.* 40:75–80.

Wolff, M.S., Lorimer, W.V., Lilis, R., and Selikoff, I.J. (1978). Blood styrene and urinary metabolites in styrene polymerisation. *Br. J. Ind. Med.* 35:318–329.

Wong, L.T., Whitehouse, L.W., Solomonraj, G., and Paul, C.J. (1980). Effect of a concomitant single dose of ethanol on the hepatotoxicity and metabolism of acetaminophen in mice. *Toxiciology* 17:297.

Woolverton, W.L., and Balster, R.L. (1981). Behavioral and lethal effects of combinations of oral ethanol and inhaled 1,1-1-trichloroethane in mice. *Toxicol. Appl. Pharmacol.* 59:1–7.

Wright, N., and Prescott, L.F. (1973). Potentiation by previous drug therapy of hepatotoxicity following paracetamol overdosage. *Scot. Med. J.* 18:56.

Yang, C.S., Tu, Y.Y., Koop, D.R., and Coon, M.J. (1985). Metabolism of nitrosamines by purified rabbit liver cytochrome P-450 isozymes. *Cancer Res.* 45:1140–1145.

Zbinden, G., Bagdon, R.E., Keith, E.F., Philips, R.D., and Randall, L.O. (1961). Experimental and clinical toxicology of chlordiazepoxide (Librium). *Toxicol. Appl. Pharmacol.* 3:619–637.

INDEX

T - #0686 - 101024 - C0 - 234/176/5 - PB - 9781138557598 - Gloss Lamination